T0250880

SpringerBriefs in Applied Sciences and Technology

Safety Management

Series Editors

Eric Marsden, FonCSI, Toulouse, France

Caroline Kamaté, FonCSI, Toulouse, France

Jean Pariès, FonCSI, Toulouse, France

The *SpringerBriefs in Safety Management* present cutting-edge research results on the management of technological risks and decision-making in high-stakes settings.

Decision-making in high-hazard environments is often affected by uncertainty and ambiguity; it is characterized by trade-offs between multiple, competing objectives. Managers and regulators need conceptual tools to help them develop risk management strategies, establish appropriate compromises and justify their decisions in such ambiguous settings. This series weaves together insights from multiple scientific disciplines that shed light on these problems, including organization studies, psychology, sociology, economics, law and engineering. It explores novel topics related to safety management, anticipating operational challenges in high-hazard industries and the societal concerns associated with these activities.

These publications are by and for academics and practitioners (industry, regulators) in safety management and risk research. Relevant industry sectors include nuclear, offshore oil and gas, chemicals processing, aviation, railways, construction and healthcare. Some emphasis is placed on explaining concepts to a non-specialized audience, and the shorter format ensures a concentrated approach to the topics treated.

The *SpringerBriefs in Safety Management* series is coordinated by the Foundation for an Industrial Safety Culture (FonCSI), a public-interest research foundation based in Toulouse, France. The FonCSI funds research on industrial safety and the management of technological risks, identifies and highlights new ideas and innovative practices, and disseminates research results to all interested parties. For more information: https://www.foncsi.org/

Jean-Christophe Le Coze · Teemu Reiman

Editors

Visualising Safety, an Exploration

Drawings, Pictures, Images, Videos and Movies

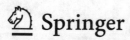 Springer

Editors
Jean-Christophe Le Coze 🆔
National Institute for the Industrial
Environment and Risks (INERIS)
Verneuil-en-Halatte, France

Teemu Reiman
Lilikoi
Lohja, Finland

ISSN 2191-530X   ISSN 2191-5318 (electronic)
SpringerBriefs in Applied Sciences and Technology
ISSN 2520-8004   ISSN 2520-8012 (electronic)
SpringerBriefs in Safety Management
ISBN 978-3-031-33785-7   ISBN 978-3-031-33786-4 (eBook)
https://doi.org/10.1007/978-3-031-33786-4

© The Editor(s) (if applicable) and The Author(s) 2023. This book is an open access publication.

Open Access This book is licensed under the terms of the Creative Commons Attribution 4.0 International License (http://creativecommons.org/licenses/by/4.0/), which permits use, sharing, adaptation, distribution and reproduction in any medium or format, as long as you give appropriate credit to the original author(s) and the source, provide a link to the Creative Commons license and indicate if changes were made.

The images or other third party material in this book are included in the book's Creative Commons license, unless indicated otherwise in a credit line to the material. If material is not included in the book's Creative Commons license and your intended use is not permitted by statutory regulation or exceeds the permitted use, you will need to obtain permission directly from the copyright holder.

The use of general descriptive names, registered names, trademarks, service marks, etc. in this publication does not imply, even in the absence of a specific statement, that such names are exempt from the relevant protective laws and regulations and therefore free for general use.

The publisher, the authors, and the editors are safe to assume that the advice and information in this book are believed to be true and accurate at the date of publication. Neither the publisher nor the authors or the editors give a warranty, expressed or implied, with respect to the material contained herein or for any errors or omissions that may have been made. The publisher remains neutral with regard to jurisdictional claims in published maps and institutional affiliations.

This Springer imprint is published by the registered company Springer Nature Switzerland AG
The registered company address is: Gewerbestrasse 11, 6330 Cham, Switzerland

Contents

Chapter 1
Visualising Safety

An Exploration

Jean-Christophe Le Coze and Teemu Reiman

Abstract Safety research and practice has struggled with how to describe, define and represent safety in order to improve understanding or to communicate its importance. Though visual representations are widely used, little research on visualisation and its impact has been undertaken. We provide a brief overview of existing work in this area, in areas including cognitive engineering and ethnography, and provide an introduction to the chapters that constitute this volume on the visualisation of safety.

Keywords Visualisation · Visual artefacts · Media · Safety · Risk

1.1 Introduction

Safety research and practice has struggled with how to describe, define and represent safety in order to understand the concept better or to communicate its importance. Visual representations have been produced since the beginning of safety practice and research, but research on visualisation has been dispersed, and rather scarce so far. Some notable exceptions are the works in cognitive engineering stemming from the need to design computer interfaces [1], in human factors when developing a better understanding and design of safety warning signs [2], in social, political and historical perspectives of safety posters [3, 4], in ethnographic studies based on socio-material sensitivities [e.g., 5, 6], in graphic design when commenting engineering decisions that led to disasters [e.g., 7, 8], in reflections on the graphic dimension of safety models [9–11], and in analysis of the safety narratives of movies [12–14]. We introduce these studies very briefly below.

The role of visualising in our understanding of safety has been little conceptualised previously. Emphasis on technical components, on actions of various individuals

J.-C. Le Coze (✉)
Ineris, Verneuil-en-Halatte, France
e-mail: jean-christophe.lecoze@ineris.fr

T. Reiman
Lilikoi, Lohja, Finland

© The Author(s) 2023
J.-C. Le Coze and T. Reiman (eds.), *Visualising Safety, an Exploration*,
SpringerBriefs in Safety Management, https://doi.org/10.1007/978-3-031-33786-4_1

through cognition, organisation or regulation thanks to established disciplines such as engineering, cognitive psychology, sociology of organisations or management and political sciences have framed our grasp of safety in the past 30 years. Little has been granted to a transversal appreciation of visual artefacts (e.g., pictures, images, videos, drawings, movies, posters) across descriptions and conceptualisations from this diversity of disciplines. Yet, in the daily practices of the high-risk industries, a world of images exists made of warning signs, diagrams (including PID: process instruments diagrams), alarms, thresholds, schematic, tables, pictograms, posters, procedures, schedules, Gantt charts, indicators, maps, logs, forms, causal diagrams but also, photographs, videos or movies supporting, guiding and providing contexts for the social fabric of safety.

1.2   Ways of Visualising

Posters and warning signs are probably the first visualisations which come to mind when thinking about safety, whether as an employee in a factory or in a high-risk system, as a user of services such as transportation, as a consumer of a diversity of products, or simply as a third party exposed to externalities of organisations. Design of warning signs and behavioural response to them by the diversity of audiences have been studied for many years now [2]. Interest in safety information displayed in other communicative art forms such as posters in factories have also been granted, exploring the views of workers and safety that they embody and their evolution in time as well as difference across countries [3].

Interfaces are also quite clearly in the mind of many when it comes to visualising because of how much they frame activities of process operators in control rooms, of pilots in cockpit, of surgeons in operating theatres, etc. This has been an important research area in the field of cognitive engineering from the 1980s, with now many established writers and standards publications on the topic [1].

Engineers also rely on drawings and visualisations to help decision-making, e.g., when identifying hazards, assessing risks or designing processes. Analysis by Tufte of the graphics which supported the decision rationale of the Challenger launch in 1986 has become a landmark study of this aspect of engineering decision-making [7]. By omitting to exhibit in an appropriate manner data which were available and that they knew to be important to ground their rationale, engineers failed to provide a more complete view of the relationship between temperatures and O-rings' problems. *"The chart makers had reached the right conclusion. They had the correct theory and they were thinking causally, but they were not displaying causally"* [7, p. 44].

Operators and engineers are of course not the only users of graphics. Managers also rely on them. The most evident example in the field of safety are the trends based on indicators which are built and followed to steer organisations' degree of achievement in preventing health, occupational or process events. The widespread use of ratios in occupational safety calculating the number of days off for injured people per hours worked and also number and magnitude of various incidents are transformed into graphics. The validity of these indicators as safety measures has been questioned by safety science, but their use in industry is still widespread. This is probably at least in part due to their easy visualisation and the apparent easiness of interpreting the visual.

Safety researchers are also great producers and users of drawings, pictures and visualisations for conceptualising the phenomena they attempt to grasp. Examples abound of drawings supporting the framing of scientific areas of investigation: human error, sociotechnical systems, comparing high-risk systems or accident causation have all been assisted by pictures, drawings or images other than texts [9]. They are designed by various authors and circulate among peers, sometimes bridging research and practice, and shaping the field through their heuristic visual properties. This drawing creativity is quite widespread among safety professionals too who are not only consumers but also designers and producers of their own drawings, pictures and visualisations that they regularly use in practice.

Some visualisations in safety science have also remained in use despite a lot of scientific evidence against the theoretical models underlying the visualisation. Examples of these are the domino model of accident causation, accident-incident triangle and the Swiss Cheese accident model. Again, their visual properties make them attractive to a general audience. Sometimes, the attractiveness of the visualisation may be a more important factor explaining the diffusion of the model than its underlying logic concerning the phenomena depicted.

This very short description of the world of pictures, drawings and visualisation does not exhaust the diversity of other image-based artefacts, such as photographs, Powerpoint, cartoons, videos, TV programmes or movies which also offer some support for descriptions, interpretations, narratives and understandings of safety for a wider audience than the people populating workplace, factories, high-risk systems or the users of services (e.g., transport).

For instance, recent movies which come to mind are blockbusters such as *Sully* or *DeepWater Horizon*, and popular programmes about aircraft crashes or other disasters are quite regularly broadcasted on TV, such as, in French, "*la minute de vérité*" (the minute of truth), which are extremely useful in human factors or sociology of safety training courses. Moreover, in the context of our increasingly digital world, safety movies available on *YouTube* or *Daily Motion*, whether from practitioners or academics (e.g., conference, courses), are also now a widespread phenomenon, which provide support for the visual diffusion of safety research, practices and ideas.

1.3 Chapters of This Book

This book is derived from a workshop held in June 2019 in Royaumont, France, to address the issues discussed above. The above introduction, accompanied by some illustrations that we were unable to reproduce here for copyright reasons, was sent to a selected group of renowned scientists in the field of safety and visualisation.

Accompanying the general introduction was description of the aim and preliminary topics for the workshop. These are reproduced below.

The aim of the workshop was to explore this realm of visualisations, images, drawings, pictures, photographs and videos in the field of safety. It wished to build a better appreciation of how these diverse artefacts contribute in their own specific way to the social fabric of reliability, safety or resilience. It was an exploratory workshop, aware of the limited number of studies available, but willing to open many different lines of investigation; a workshop therefore multidisciplinary in its prospect. It wished to increase our awareness of the incredible complexity of the current sociomaterial dynamics of our mediatised, digitalised and globalised world. *"Like it or not, the emerging global society is visual"* [15, p. 4].

The following non-exhaustive list of questions to explore was included in the original call:

- Which are the examples of successful visualisations in safety, in research, in practice? Do we know why?
- How do safety pictures, signs, drawings, visualisations or videos in safety have evolved over time? Can we characterise this evolution?
- How to classify the diversity of visual artefacts in safety? In relation to what properties? From simplicity to complexity? From dynamic to static? To what features?
- What is your experience as a researcher of drawings, pictures or videos? How important are they to your research process, from theorising to communicating? Are visualisations only appropriate to communicate about safety, or also to conceptualise safety?
- How do pictures, drawings or videos contribute to the enactment of safety? What do we know about the effects/influence of visualisation on safety (structures, processes, outcome)? What kind of agency do pictures, drawings or visualisations have in safety? How performative are they? How can we describe conceptualise or even measure this performativity?
- How do practitioners and/or researchers produce, use and disseminate diverse visualisations in their daily activity? How important is it to their coordination, cooperation or communication?
- How does visualising a concept change the concept? Do visualisations complement, accompany or replace texts? Are safety models and theories best conveyed by drawings than texts?
- How do research drawings, pictures or videos contribute to our framing of safety as a scientific object?
- How do videos or movies portray safety through their narratives? Can they be useful support for safety management, for public awareness or dread? How?
- What kind of opportunities new technology offers in the context of safety knowledge production, transmission and use?
- How does big data shape new need for visualisations in the field of safety? Can the tools and methodologies develop in the context of big data transferred in a safety context?

- What are the limits of visualising? Are there some phenomena that cannot or should not be visualised? Are there dangers in visualising complex phenomena such as safety? How do you visualise risk and uncertainty?
- How do visualisations guide the attention of public and experts on risks and safety? Can a vivid visualisation create biases (e.g., availability bias) that distract attention from other, less visualised, risks or types of safety?

1.4 Organising the Workshops and the Book

The workshop was organised by an international, interdisciplinary study group New Technologies and Work (NeTWork). It was the 34th workshop by NeTWork, which has been active since the 1980s (network-network.org).

The concept of these workshops has been to maximise an international interdisciplinary discussion on topics of technology, work and management. Therefore, only a small group of researchers and practitioners is invited to each workshop. The group is led by an international core group of scientists who evaluate the former workshops and plan the forthcoming. The core group is responsible for the topic as well as for the invitation to a limited group of experts invited, *ad personam*, in the field of the workshop's interest.

The call was distributed by the core group of the NeTWork with the aim of inviting experts familiar with the topic of the call. As is the tradition in the NeTWork workshops, each participant was asked to write a "position paper" summarising the key points of their contribution and submit it to the organisers a few weeks before the workshop. Position papers were shared among the participants prior to the workshop, and each participant was expected to read them. This contributed to active participation and discussion during the three day workshop.

Chapters of this book are based on the position papers presented and discussed at length during the workshop and revised based on the feedback received. The "Brief" nature of the collection means that contributions are limited in length. The book's open access licence made it difficult to obtain reprinting rights to several legacy documents including many visualisations included in the original position papers. A positive side of open access is that authors can reuse and develop their ideas further, and also reach a wider audience. Chapters in this book thus contain the essence of the arguments but omit much of the background explanation and accompanying "legacy" visualisations. This is an interesting observation as such concerning the use and reuse of visual artefacts in safety science research.

In Chap. 2, Paul Swuste et al. take a look at the history of drawings, posters and photos in safety and safety science. Their position paper included numerous images that captured the development of visual side of safety through the years. The chapter in this book includes their core arguments but omits much of the visualisations. Their chapter illustrates how the early visualisations show a clear message of fear and guilt, whereas organisational factors also appear in more recent visualisations.

Chapter 3 by Aurelien Portelli et al. addresses a particular case of educating nuclear workers through images by looking at the work of Jacques Castan. He was an illustrator at the French Commissariat à l'énergie atomique (CEA) and was in charge of illustrating radiation protection campaigns in the Marcoule nuclear site during the 1960s. Portelli et al. focus on a series of posters designed by Castan on the topic of dosimeter films and pens. They identify three main iconographic elements within the series: anxiety, anthropomorphism and sublimation.

In Chap. 4, Patrick Waterson tackles the ways of seeing and not seeing safety by looking at safety models and their associated visualisations. Drawing on the path-breaking contribution of John Berger, he contemplates why visual representations are so common in safety models and what does the use of visual representations tell us about research and practice in safety science.

In Chap. 5, Torgeir Haavik discusses visualisation and representations. He looks at representation as immutable mobiles, borrowing the concept from Bruno Latour to show what makes them so powerful tools in safety. Haavik uses the sharp end/ blunt end metaphor as a case for illustrating the challenges of immutable mobiles, for example when travelling across contexts and scales.

Chapter 6 by Erik Hollnagel, based on his extensive production of safety-related visualisations, asks the fundamental question of whether safety as such can be visualised. He makes a distinction between visualisation *for* safety and visualisation *of* safety. The latter Hollnagel deems impossible, whereas the former can be achieved, depending on the purpose of the visualisation. He suggests that visualising should not be a purpose in itself, but a means to achieve a purpose.

In Chap. 7, Doug Smith and colleagues move the reader into the animated, digitalised and software-oriented side of visualisation, and demonstrate how functional signatures can be used to visualise complex industrial operations with the help of computers. Functional signatures are an extension of the functional resonance analysis method (FRAM) that can help monitor complex operations and improve tractability. They provide an example of the functional signature concept based on ship navigation in a simulated ship environment.

Chapter 8 by John Flach offers a control theoretic perspective on safety and visualisation. He presents the design principles of semantic mapping and systematicity and argues that these are fundamental to all forms of representations. Flach demonstrates how the way we are able to visualise the state of the system is central to our ability to anticipate, and control, risk.

In Chap. 9, Charles Stoessel and Raluca Ciobanu present a safety design approach to occupation safety in revamping operations. Their chapter takes a look at how design engineers perceive specific design safety issues pertaining to the revamping of existing facilities. The authors discuss the formation of situated safety skills and use of visualisation tools to improve safety through design.

Chapter 10 by Kaupo Viitanen and Teemu Reiman describe how a network visualisation method was developed and used in supply chain quality and safety assurance of a nuclear power plant construction project which relies on a myriad of contractors and subcontractors. The method was developed as a solution to better make sense of how a project network of multiple organisations creates preconditions for safety.

In Chap. 11, Gisquet and Rot show with an ethnographic case study how visualisation helps to maintain safety requirements in construction of underground infrastructures, an extension of the Paris metro in particular. They show how visual artefacts such as maps, notes, visual plans and schedules and other visual aids advance safety by helping participants inhabit, discuss and synchronise their workspaces.

Chapter 12 by Shane Dixon and Tim Gawley examines the film *Only the Brave* (2017), which recounts the real story of the deaths of 19 wildland firefighters in the 2013 Yarnell Hill Fire in Arizona, USA. They illustrate how film can visually communicate the story of an accident and how narrative choices affect which factors of the incident are highlighted and which are excluded.

In Chap. 13, we conclude and provide some future research directions.

References

1. K.B. Bennett, J.M. Flach, *Display and Interface Design* (CRC Press, 2011)
2. K.R. Laughery, M.S. Wogalter, A three-stage model summarizes product warning and environmental sign research. Safety Sci. **61**, 3–10 (2014)
3. A. Menendez-Navarro, *The Art of Preventive Health and Safety in Europe* (European Trade Union Institute, 2015)
4. S. Travadel, A. Portelli, C. Parizel, F. Guarnieri, Les figures de l'infime, la radioprotection en images. Techn. Culture **2**(68), 110–129 (2017)
5. T. Haavik, Remoteness and sensework in harsh environments. Saf. Sci. **95**, 150–158 (2017)
6. S. Tillement, J. Hayes, Maintenance schedules as boundary objects for improved organizational reliability. Cogn. Technol. Work (2018)
7. E.R. Tufte, *Visual Explanations. Images and Quantities, Evidence and Narratives* (Graphics Press, Cheshire, CT, 1997)
8. E.R. Tufte, *Beautiful Evidence* (Graphics Press, Cheshire, CT, 2006)
9. J.C. Le Coze, Visualising safety, in *Safety Research: Evolutions, Challenges and New Directions*, eds. by J.C. Le Coze (CRC Press, Boca Raton, FL, 2019)
10. J.C. Le Coze, New models for new times An anti-dualist move. Safety Sci. **59**, 200–218 (2013)
11. P. Swuste, Is big data risk assessment a novelty? Safety Reliabil. **3**, 134–152 (2016)
12. S. Dixon, T. Gawley, Crude exploration. Portraying industrial disaster in Deepwater Horizon, a film directed by Peter Berg, 2016. New Solutions J. Environ. Occup. Health Policy (2017)
13. H. Laroche, De "Sully" à Trump: la légende de l'individu contre l'organisation. The Conversation. 9 Dec 2016
14. A. Portelli, F. Guanieri, C. Martin, Le nucléaire fait son cinéma. Revue Générale du Nucléaire **1**, 96–101 (2014)
15. N. Mirzoeff, *How to See the World. An Introduction to Images, From Self-Portaits to Selfies, Maps and Movies, and More* (Basic Books, London, 2016)

Open Access This chapter is licensed under the terms of the Creative Commons Attribution 4.0 International License (http://creativecommons.org/licenses/by/4.0/), which permits use, sharing, adaptation, distribution and reproduction in any medium or format, as long as you give appropriate credit to the original author(s) and the source, provide a link to the Creative Commons license and indicate if changes were made.

The images or other third party material in this chapter are included in the chapter's Creative Commons license, unless indicated otherwise in a credit line to the material. If material is not included in the chapter's Creative Commons license and your intended use is not permitted by statutory regulation or exceeds the permitted use, you will need to obtain permission directly from the copyright holder.

Chapter 2
Drawings, Posters and Metaphors in Safety Science: Some Historical Remarks

Paul Swuste, Peter Schmitz, Karolien van Nunen, and Genserik Reniers

Abstract Safety visualisations and their influences on safety concepts are presented. Visualisations like safety posters show a clear message of fear and guilt. This changes after World War II, due to a more tolerant atmosphere. Latent, organisational factors as decisive elements of accident processes appear in visualisations. An example shows a method to follow accident scenarios in real time.

Keywords Historical analysis · Accident prevention · Metaphors

2.1 Introduction

One of the first examples of a visualisation in Western European literature came from Agricola's standard work '*De Re Metallica*'. This book on geology, mineralogy and mining devoted a few pages to accidents of miners. Foreman of mines should anchor ladders in mine tunnels to prevent sliding of these ladders and fall-accidents of miners (Fig. 2.1). Mine entrances should not face the North. In winter times with freezing northern winds, miners could lose their grip [1].

From the late nineteenth-century posters became popular, promoting public health and warned against excessive use of alcohol, tuberculosis, syphilis. Safety posters were produced after World War I. Also, safety concepts, theories, models and metaphors were visualised. This article gives some examples of these visualisations, based upon recently published historical overviews of developments in

P. Swuste (✉) · P. Schmitz · K. van Nunen · G. Reniers
Safety Science Group, Delft University of Technology, Delft, The Netherlands
e-mail: paul@paulswuste.nl

P. Schmitz
OCI Nitrogen, Geleen, The Netherlands

G. Reniers
Antwerp Research Group on Safety and Security (ARGoSS), University of Antwerp, Antwerb, Belgium

© The Author(s) 2023
J.-C. Le Coze and T. Reiman (eds.), *Visualising Safety, an Exploration*,
SpringerBriefs in Safety Management, https://doi.org/10.1007/978-3-031-33786-4_2

Fig. 2.1 Sixteenth-century
mine shafts

the safety domain in a *Safety Science* series on safety knowledge development
2009–2020 [18–25]. Questions below are leading for this contribution:

- Is there a trend in visual presentations and how can this development be
  characterised?
- Are visualisations changing safety concepts?
- What are limitations of visualisations?

2.2 Safety Posters

Figure 2.2 shows one of the first US safety posters and the first Dutch safety poster.
The Safety First Movement, which started in the US, is dominant at that time, aiming
at safe behaviour of workers and at hazards of moving parts of machines: safety
technique.

Safety technique, the enclosure of rotating parts of machines and the fencing of
heights started in 1844 in the UK and spread over Europe and the US. US authors
publish many practical examples of safety techniques in various industrial sectors.
The posters refer to behavioural consequences of unguarded rotating machinery.
National figures of accidents are available in the US in 1907. It shows a very high
death toll in the American steel industry (0.6–0.8 per 100 man-hour), compared to
Germany (0.2 per 100 man-hour) [2, 3]. The reliability of these figures is not clear, but
the high occupational mortality in the American industry, compared to Europe, was

Fig. 2.2 Left: Dutch safety poster from 1922 (why did you not wear a cap like me? Loose hair in proximity of machines and gears is dangerous). Right: US safety poster from 1919

a repeated argument in various publications. US Steel, the largest steel company, started in 1906 the '*Safety First Movement*' [15]. A 1913 poster on occupational accidents shows ignorance, indifference, carelessness as main causes of accidents (Fig. 2.3).

Fig. 2.3 US Safety First Movement poster, 1913

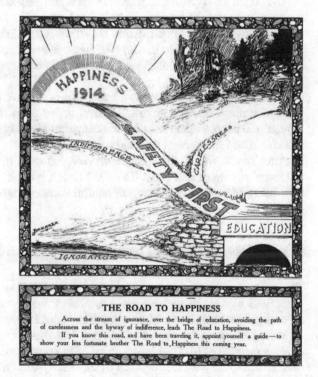

Fig. 2.4 Dutch safety poster from 1922: because he was careful, Grandpa became 70 without suffering an accident

The Dutch Safety Museum started in 1893 and played an active role in promoting occupational safety. It published a monthly journal *The Safety Journal* in the 1920s and had a weekly radio presentation on safety-related issues. The safety posters of this institute had a simple message of hazards and family values [11] (Fig. 2.4). Religious parties dominated Dutch politics at that time and posters avoided any political message.

For fear of tensions in companies, there is no reference to class differences, or shortcomings of management. Unions, active from the late nineteenth century onward, did not pay much attention to occupational safety. Their topics were salaries, working hours and general working conditions.

After World War II, the political climate changed. In the 1970s, 'humanisation of labour' became a new topic. The 1980 Dutch Working Conditions Law included well-being of workers. Posters were neutral without a moral undertone (Fig. 2.4).

2.3 Safety Concepts, Theories, Models and Metaphors

The concept of an accident as a process was published in the 1920s [5]. DeBlois postulated hazards being equivalent to energy, and process disruptions as causes of occupational accidents. Management decisions were the centre of the process. 15 years later, Heinrich presented his first visualisation of an accident process, the domino metaphor [10]. The '*unsafe act*' of the victim was the centre of the domino metaphor, in line with the message of the Safety First Movement. The strength of

Fig. 2.5 Death calendar showing the number of work-related deaths by day, Allegheny County, Pennsylvania

this metaphor is its simplicity. Even today the metaphor has a major impact on safety professionals. Another publication before DeBlois also showed external factors as causes of accidents. In 1910, Crystal Eastman [6] published a death calendar. In the period of one year, the mortality amongst steelworkers in the US Steel plant in Allegheny district, Pittsburgh was 526 (Fig. 2.5). These numbers were staggering. Every day one and a half fatal accidents occurred. Her research was a first attempt of a socio-technical approach to safety. According to Eastman, causes of the accidents were the fatal interactions and uneducated employees, mostly kids, send by managers to dangerous machines.

Different scientific disciplines are active in safety research. Engineers look at accident processes, hazards, scenarios and barriers. Organisational processes and decision-making in companies is the focus of sociologists, while behaviour is a starting point of psychologists. A special group of engineers, risk analysts, calculate risks of major accidents using failure probabilities of technical components and probabilities of consequences, mostly mortality. Apart from risk analysis, all other disciplines look for factors which bring production system into an uncontrolled state. The terminology differed over time and discipline. The sociologist Turner, promoting a socio-technical approach to safety, defined an '*incubation period*' of major accidents, a period of systematic risk denial of an organisation [28]. The engineer Kjellén defined causes of an occupational accident processes as '*process disturbances*' and '*loss of control*' [12].

Different models and metaphors were published in the 1990s; the 'Tripod' model of the psychologist Groeneweg, named after a three-legged dog seen during field-work, the metaphors '*Swiss cheese*', and the '*bowtie*'. Operational disturbances, with incomplete barriers, caused by decision-makers were starting points of accident processes, both of occupational and major ones [7]. The model referred to the 'hazard-barrier-target' model from the 1960s [8]. The 11 'Basic Risk factors' (BRF) were mostly organisational factors. These factors were a specification of Turner's incubation period. The accident process started with decision makers, like the 'blunt end' of the well-known Swiss cheese metaphor of the psychologist Reason [17].

The bowtie was an engineering metaphor for both occupational and major accidents, without unsafe acts, or psychological precursors. There are multiple scenarios (arrows from left to right), barriers (the rectangular shapes in the scenarios) and a *'central event'*. This is the centre of the accident processes when hazard(s) became uncontrollable [28]. There were no holes in barriers. Instead of managerial factors, the upward lines determined the quality of barriers, like the BRF's in Tripod. The metaphor had different time dimensions. Deficient or absent barriers and management factors had an effect over a long period of time, like Turner's incubation period. Accident scenarios left of the central event could take week, or longer to develop. If a central event became active, scenarios to consequences would unroll very quickly.

A conceptual model of safety was Rasmussen's 'Drift to danger' [16]. The financial dominance of *'the market'* initiated a management focus of cost-effectiveness of production resulting in an increased pressure on workers. According to Rasmussen, human failures were not causes of major accidents, but systematic migration of organisational behaviour towards an accident was.

(Re)design is the topic of engineers. In the 1980s, the concept of *'Inherently safe design'* was published [13]. 'Small is better', and the use of safe, less toxic and less flammable chemicals was his message. This concept was strong because of its simplicity. Kletz proposed transparency, because morally it was preferably to inform society 'if we know, we must tell'.

LOPA (layers of protection) was another design concept for the process industry, developed in the same period as inherent safe design [4]. This design strategy followed a *'defence in depth'* principle. LOPA implied multiple layers of independent safety barriers for the mechanical integrity of a production system, to prevent emission or loss of containment. But all barriers have their weaknesses, like the resident pathogens of Swiss cheese. When operators were unaware of failures in one of more layers, an unnoticed scenario developed after a process failure, a so-called wildness in the wait. According to Rasmussen, there was a *'fallacy of defence in depth'*.

Monitoring accident scenarios is possible when management factors, the latent factors, are linked to process indicators in a bowtie analysis. These factors can influence hazards (inherently safe design), scenarios (training, information) or technical barriers for specific scenarios. Scenario-specific interventions are possible and puts safety closer to the core business of the company. Recent research in a manufacturing company and a chemical plant showed these possibilities.

Accidents with pallet movers for internal transport were a major problem in a Belgium manufacturing company. Literature research, interviews and workplace observations gave an overview of actual and possible accident scenarios. Leaking cubitainers, resulting in spills of products on floors, increasing breaking distance of

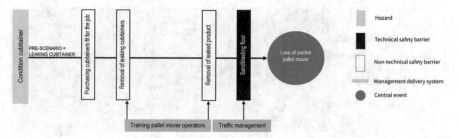

Fig. 2.6 An elaboration of the left side of a bowtie

pallet movers and destabilises loads was a major issue in the company. The bowtie analysis showed management factors with an impact on both technical and non-technical safety barriers. Training operators is a non-technical barrier, and only effective when training concentrates on possible and occurring scenarios, barriers and central events. The next step was to define process indicators for management and workers to follow the development of scenarios [14, 26]. Figure 2.6 shows the influence of different management factors on elements of the bowtie. Sandblasting floors was a technical barrier which prevented pallet movers from slipping.

Major accidents in chemical plants are complex. Following a similar strategy as above, the bowtie analysis of a NH_3 producing plant resulted in a visualisation of scenarios. Cracking of natural gas (CH_4) in the Meka 1 is the first step of the process. A vertical tube reactor and a secondary reformer produce hydrogen (H_2) under high pressure (40 bar) and temperature (1000 °C).

Figure 2.7 shows the secondary reformer. Cooling of the secondary reformer (R3201) with water jackets is essential, due to process conditions. When these water jackets fail, excessive heat exposes and weakens the reformer wall with possible catastrophic consequences and a massive emission of highly flammable and explosive gas. The activation of alarms on low/high temperature, low/high flow, and low level is connected to accident scenarios, coming from the literature, company documentation, and interviews with operators, maintenance and safety staff, technical and operational managers and the CEO of the company [26].

Various levels of information are presented:

- Level 1 shows the production steps at the NH_3 production. This level provides an overview of the whole plant, relevant for a CEO. The red dot in the level 1 dashboard shows a problem at Meka 1.
- Level 2 shows the installations of Meka 1, relevant for the safety, and plant manager. The second reformer R3102 is red, indicating the location of the problem.
- Level 3 (see Fig. 2.8) shows the second reformer's scenarios, including instrumental safeguards and barriers. This level gives information on the operational status of individual barriers, relevant for operators, and maintenance, safety, and mechanical engineers. The first scenario, water jacket failure, shows three alarms. FIAL1110 (flow indicator alarm, low flow) is red, this alarm is activated, and two

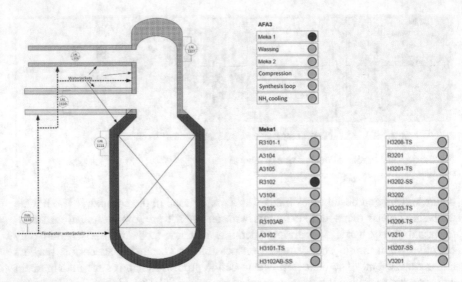

Fig. 2.7 Reformer R3102 of Meka 1 (left), with the level 1 (top right) and level 2 (bottom right) dashboards

Fig. 2.8 Level 3 indicators and scenarios for reformer 2

LAL's (level alarm, low level) are yellow. These alarms are not reliable. A scenario related to overheating of the second reformer starts to unroll.

2.4 Discussion and Conclusion

Posters show a major development over time. They reflect the political atmosphere and dominant views on accident processes. In the 1920s, fear was the central message. From the 1970s onwards, the message was non-moralistic. In the early days, some authors pointed at external factors, and managerial influences of accidents, but the dominant cause attributed was unsafe acts of workers. Posters were popular, and cheap, and decorated walls of factories. They showed a safety interest of a safety department, or a vision of management. It is questionable whether posters had any influence on accident processes. No research was conducted to give credit to their effects.

Between the 1970s and 1990s, a very productive period of safety science, theories, models and metaphors of accident processes showed a growing focus on organisational factors. Internal, or latent factors in combination with external factors created unstable production systems, leading both to occupational accidents and major accidents with a high media coverage. Graphical presentations contained arrows. The arrow in Tripod, as in Swiss cheese, might refer to correlations, to causal relations, or even to accident scenarios. Most likely this reflected a social science interpretation of accident processes with less focus on hazards and scenarios. The bowtie metaphor depicted specific scenarios leading to loss of control. This metaphor offered opportunities to follow the scenarios in real time, offering management and workers an overview of the safety state of their processes. A clear relation between management and workers' activities and (major) accident processes is vital, because (major) accidents are not prevented by regulatory compliance, nor by ISO standards.

Risk management, calculating frequencies and probabilities of major accidents, and cost benefits of safety measures are an essential part of a management approach. But rational arguments and quantifications only have a limited influence in decision-making. The formal rationality of organisations is doubtful. Management actions are rarely preceded by a comprehensive problem analysis, or an overview of possible actions. Generally, Rasmussen's external factors, such as 'the market', play a dominant role. Rationality is more a façade, and reality is like a metaphor for how people in an organisation understand the flow of information they have to deal with [29]. It is questionable whether managers are always interested in quantitative information. Having witnessed a major accident, or reputational arguments will often guide their safety initiatives [9].

Theories, models, and metaphors in this paper also present a rational explanation of accident processes. This knowledge only partly enters the domain of safety professionals, process engineers or corporate management. Some concepts are generally accepted, like dominoes, Swiss cheese, Normal Accidents, High Reliability Organisations, Inherently safe design and LOPA. Other concepts as Tripod, bowtie, process disturbances only have a local, or national appreciation. Barry Turner's theory from 1978 was a special case, staying dormant for almost 20 years. His concept of 'incubation', risk blindness of organisations, is important and later transformed as latent factors (Swiss cheese), basic risk factors (Tripod), and management factors (bowtie). One explanation for its dormant state was that Turner's article [27], published in management journals, was rarely read by safety scientists.

It is not clear why safety concepts gain acceptance by safety professionals and beyond. Maybe it is language, simplicity of concepts or effective promotion by authors. The dominant status of the domino metaphor is directly related to Heinrich's position. His production of easy-to-understand concepts and ratios is impressive. Also, James Reason was effective in promoting his cheese metaphor. But process disturbances, as postulated by Eastman, DeBlois and Kjellén got less attention. Maybe these concepts are too complex to communicate or not promoted externally.

References

1. G. Agricola, *De Re Metallica* (H. Hoover, L. Hoover, Trans., 1950) (Dover Publications, New York, 1556)
2. M. Aldrich, *Safety First: Technology, Labour and Business in the Building of American Safety 1870–1939* (John Hopkins University Press, Baltimore, 1997)
3. Anonymous, Industrial accident statistics. Science **42**(1077), 238–239 (1915)
4. CCPS, Center for Chemical Process Safety, simplified process risk assessment. American Institute of Chemical Engineers, New York (2001)
5. L. DeBlois, *Industrial Safety Organization for Executives and Engineer* (McGraw-Hill Book Company, New York, 1926)
6. C. Eastman, Work-accidents and the law. The Pittsburgh survey. Charities Publications Committee, New York (1910)
7. J. Groeneweg, *Controlling the Controllable, the Management of Safety* (Proefschrift Rijksuniversiteit Leiden, DSWO Press, Leiden, 1992)
8. W. Haddon, A note concerning accident theory and research with special reference to motor vehicle accidents. Ann. N. Y. Acad. Sci. **107**, 635–646 (1963)
9. A. Hale, Management of industrial safety, in *Encyclopaedia of Life Support Systems* (UNESCO, 2004)
10. H. Heinrich, *Industrial Accident Prevention, a Scientific Approach*, 2nd edn. (McGraw-Hill Book Company, London, 1941)
11. H. Hermans, Een monster loert... De collectie historische gezondheidsaffiches van de Universiteit van Amsterdam (A monster lurks. The collection of historical health posters from the University of Amsterdam) (Vossiuspers, Amsterdam, 2007)
12. U. Kjellén, The role of deviations in accident causation. J. Occup. Accid. **6**, 117–126 (1984)
13. T. Kletz, *Cheaper, Safer Plants* (Institute of Chemical Engineers, Rugby, 1984)
14. K. Nunen, P. van Swuste, G. Reniers, N. Paltrinieri, O. Aneziris, K. Ponnet, Improving pallet mover safety in the manufacturing industry. A bowtie analysis of accident scenarios. Materials **11**(1955), 1–19 (2018)
15. L. Palmer, The history of the safety movement. Ann. Am. Acad. Polit. Soc. Sci. **123**(1), 9–19 (1926)
16. J. Rasmussen, Risk management in a dynamic society: a modelling problem. Saf. Sci. **27**(2–3), 183–213 (1997)
17. J. Reason, *Managing the Risks of Organizational Accidents* (Ashgate, Aldershot, 1997)
18. C. Gulijk, P. van Swuste, W. Zwaard, Ontwikkeling van veiligheidskunde in het interbellum en de bijdrage van Heinrich (Safety during the interbellum, Heinrichs' contribution). Tijdschr. Toegepaste. Arbowetenschap. **22**(3), 80–95 (2009)
19. P. Swuste, C. Gulijk, W. van Zwaard, Safety metaphors and theories, a review of the occupational safety literature of the US UK and The Netherlands till the first part of the 20th century. Saf. Sci. **48**, 1000–1018 (2010)
20. P. Swuste, C. Gulijk, W. van Zwaard, Y. Oostendorp, Occupational safety theories, models, and metaphors in three decades after WO II, in the United States, Britain, and The Netherlands. Saf. Sci. **62**, 16–27 (2014)
21. Y. Oostendorp, S. Lemkowitz, W. Zwaard, C. Gulijk, P. van Swuste, Introduction of the concept of risk in The Netherlands. Saf. Sci. **86**, 205–219 (2016)
22. P. Swuste, C. Gulijk, W. van Zwaard, S. Lemkowitz, Y. Oostendorp, J. Groeneweg, Developments in the safety science domain, in the fields of general and safety management between 1970–1979, the year of the near disaster at Three Mile Island, a literature review. Saf. Sci. **86**, 10–26 (2016)
23. P. Swuste, J. Groeneweg, C. Gulijk, W. van Zwaard, S. Lemkowitz, Safety management systems from Three Mile Island to Piper Alpha, a review in English and Dutch literature for the period 1979 to 1988. Saf. Sci. **107**, 224–244 (2019)

24. P. Swuste, C. Gulijk, J. van Groeneweg, F. Guldenmund, W. Zwaard, S. Lemkowitz, Development of safety management between 1988–2010 (occupational safety). Review of safety literature in English and Dutch language scientific literature. Saf. Sci. **121**, 303–318 (2020)
25. P. Swuste, C. Gulijk, J. van Groeneweg, W. Zwaard, S. Lemkowitz, From Clapham junction to Macondo, deepwater horizon: risk and safety management in high-tech-high-hazard sectors a review of English and Dutch literature: 1988–2010. Saf. Sci. **121**, 249–282 (2020)
26. P. Swuste, J. Theunissen, P. Schmitz, G. Reniers, P. Blokland, Process safety indicators. J. Loss Prev. Process. Ind. **40**, 162–173 (2016)
27. B. Turner, *Man-Made Disasters* (Butterworth-Heinemann, Oxford, 1978)
28. K. Visser, Developments in HSE management in oil and gas exploration and production, in *Safety Management, the Challenge of Change*. ed. by A. Hale, M. Baram (Pergamon, Amsterdam, 1998), pp.43–66
29. K. Weick, K. Sutcliff, Managing the unexpected, in *Resilient Performance in the Age of Uncertainty*, 2nd edn. (Wiley, Chichester, 2001)

Open Access This chapter is licensed under the terms of the Creative Commons Attribution 4.0 International License (http://creativecommons.org/licenses/by/4.0/), which permits use, sharing, adaptation, distribution and reproduction in any medium or format, as long as you give appropriate credit to the original author(s) and the source, provide a link to the Creative Commons license and indicate if changes were made.

The images or other third party material in this chapter are included in the chapter's Creative Commons license, unless indicated otherwise in a credit line to the material. If material is not included in the chapter's Creative Commons license and your intended use is not permitted by statutory regulation or exceeds the permitted use, you will need to obtain permission directly from the copyright holder.

Chapter 3
Educating Nuclear Workers Through Images: The Work of Jacques Castan, Illustrator of Radiation Protection in the 1960s

Aurélien Portelli, Frédérick Lamare, Sébastien Travadel, and Franck Guarnieri

Abstract In France, the first industrial-scale nuclear reactors were built by the French Atomic Energy Commission at Marcoule during the fifties. Most of the staff who were recruited at the time knew nothing about such risks, and their inexperience made it difficult to protect them. In response, the Radiation Protection Service (SPR) developed a worker education programme. Its implementation drew upon the artistic talents of Jacques Castan, a draftsman of the SPR. This study highlights its contribution to worker education and showcases how its illustrations have captured the imaginary of the radiation protection. The focus on a series of posters dedicated to dosimetry devices identifies three elements—anxiety, anthropomorphism, sublimation—which represent an ambiguous relationship to radioactive risk. Such ambiguity can be compared to Girard's definition of the "sacred".

Keywords Radiation protection · Education · Posters · Imaginary · Sacred

3.1 Introduction

The French Atomic Energy Commission (CEA) constructed, during the fifties, the country's first generation of industrial-scale nuclear reactors at Marcoule. In terms of safety, the CEA then faced unprecedented challenges, as workers had to be protected from increasing quantities of radioactive materials. This task was the responsibility of the Radiation Protection Service [Service de Protection contre les Radiations (SPR)], which was also responsible for educating operators about the risks of radiation. The

A. Portelli (✉) · S. Travadel · F. Guarnieri
Centre de recherche sur les Risques et les Crises (CRC), MINES Paris—PSL, Paris, France
e-mail: aurelien.portelli@minesparis.psl.eu

F. Lamare
Centre de Marcoule, Commissariat à L'Energie Atomique et aux Energies Alternatives (CEA)—CEA/Marcoule, Chusclan, France

© The Author(s) 2023
J.-C. Le Coze and T. Reiman (eds.), *Visualising Safety, an Exploration*, SpringerBriefs in Safety Management, https://doi.org/10.1007/978-3-031-33786-4_3

implementation of the latter programme benefited from the artistic talents of Jacques Castan, an SPR draftsman.

This study highlights its contribution to worker education and showcases how its illustrations have captured the imaginary of the radiation protection. In the first part, we discuss SPR doctrine, implemented by the education programme. In the second part, we present Castan's body of work. In the last part, we focus on a series of posters dedicated to dosimeters. Although the effectiveness of prevention posters has been discussed in the literature [4, 9, 17, 19], the lack of studies conducted in the 1960s on how Castan's posters were used makes it difficult to comment on their specific effect. However, the analysis identifies iconographic elements, making it possible to qualify the relationship that operators maintain with radioactive risk.

3.2 A Radioactive Risks Education Programme

While it did not invent radiation protection, the SPR rationalised and industrialised it.[1] Its success would prove to be crucial for the future of the nuclear industry. Close monitoring of radiation was not only a matter of health, but also a *sine qua non* in controlling the massive forces unleashed by science that engineers were required to control in order to produce energy.

3.2.1 Radiation Protection Doctrine

The challenge for the SPR was to discover the risks as operations unfolded and to implement prevention measures. These measures were based on the classification of workplaces according to their level of radioactive risk; the use of fixed on-site detectors; routine checks carried out by radiation protection officers; and the distribution of individual protection equipment and detectors to personnel. The department was also responsible for the decontamination of equipment and clothing. Finally, it monitored radioactivity levels in effluents released by the facilities.

Over the years, the SPR developed its own doctrine for radiation protection in industrial environments, the first of which was formalised in 1965 in the *General Radiation Protection Instructions* [12]. This manual was distributed to other CEA centres and used as a model to standardise radiation protection practices and help to establish a shared culture in the emerging nuclear industry.

---

[1] The CEA created the first SPR at the Fontenay-aux-Roses site in 1951 [18]. In 1956, the Fontenay SPR was divided into two departments: the Atomic Hygiene and Radiopathology Department, and the Radiation Control and Radioactive Engineering Department. The latter's director helped to draw up the CEA's theoretical foundations regarding radiation protection [5].

Prevention also took the form of a radiation protection educational programme, which began to be developed in 1959. The SPR considered education to encompass both workers in the nuclear industry and the general public. Moreover, given that workers were recruited from the general public, the latter's reservations, if not addressed, would have hindered the expansion of the sector. "The general public will therefore have to be the subject of a general information program. On the other hand, education must be specialised when it is aimed at workers or officials in charge of radiation protection" [15]. This belief, widespread within the SPR, was part of a process of the mass publication of articles.

3.2.2 Educating Workers and the General Public

The SPR was particularly concerned that new workers failed fully to appreciate the risks of radiation. Some were overly cautious, while others took unnecessary chances. Consequently, courses were organised to demonstrate to the former that fear was not a good way to protect themselves, and show the latter why it was important to follow instructions. These events were an opportunity to explain the risks associated with the handling of radioactive materials and demonstrate how to protect against them. Instructions were illustrated by slides and videos.

With respect to the general public, the SPR organised guided tours of Marcoule, participated in educational film projects and regional exhibitions. However, the service was confronted with the problem of how to represent risks that could not be detected by the human senses. Castan's creative skills would become a key asset in meeting these educational objectives.

Born in 1929, Castan began drawing as a child and joined an architectural firm, where he trained as a draftsman and designer. In 1957, he was hired as a draftsman for the SPR. His first project was a waste pit, but it was not long before the head of the SPR noted his skills with a pencil and suggested that he illustrate Marcoule's prevention campaigns.[2]

3.3 How to Draw an Invisible Risk?

Castan immersed himself in the site's activities and learned about physics. He frequently interacted with SPR staff in the field and observed operations in workshops and laboratories. His immersion in the language of engineers and technicians helped him to capture complex technical notions, which he tried to translate into a

---

[2] In 1968, Castan stopped drawing for the SPR and became a facilitator for Marcoule's training department, before taking over as its director in 1974. He retired in 1991 and died in 2014.

more readily understood form. His creations are a testament to the golden age of nuclear energy and plunge us into the striking universe of the *mystique of Marcoule* [7].

3.3.1 Jacques Castan's Body of Work

From 1959 onwards, Castan's posters were designed to illustrate radiation protection instructions. The original drawings were made using light pencil on Bristol board. Colours were added by positioning each shade on a transparent film superimposed on the black line. The first series used offset printing but, very quickly, screen printing was adopted given its ability to perfectly reproduce the solid surfaces drawn by Castan. His corpus includes 87 posters in A3 format. Initially prepared for the Marcoule centre, from the beginning of the 1960s they were distributed to other CEA centres. Castan used humour and a multitude of striking analogies. His translation of SPR doctrine contained a world populated by characters from disparate cultural universes. His illustrated leaflets on the principles and regulations of radiation protection, such as *The Use of Dosimeter Films and Pens* (1962), share the same graphical world as his posters and take a non-brutal approach to risk prevention.

In 1960, Castan designed the comic strip *Sophie and Bruno in the land of the atom*, which tells the story of two children who visit Marcoule. The images emphasise the power of atomic energy and show futuristic installations—a testament to France's technological influence [13]. This flattering presentation of the centre's activities was designed to serve the purposes of the CEA, whose mission was to support the expansion of the sector and ensure France's economic development and energy independence [3].

In 1962, Castan performs a mural in the stairwell of the SPR building. On the one hand, the work was intended to educate workers about the activities of the SPR [14]. On the other hand, it was seen by all visitors to Marcoule. The centre's activities were a source of concern for the surrounding population. The image responded to these concerns, showing that the risks were under control. This tranquil scene was intended to reassure the viewer and make nuclear power a socially desirable industry.

More modest than his mural, in 1966 Castan created a board game entitled *The noble game of the laws of radio protection*. The central part defines the rules of the game. The circle of boxes reproduces elements taken from the comic strip, posters and leaflets. The game can thus be seen as a synthesis of his earlier creations. Understood as a mediator between workers and their families, it aims to make radiation protection a soothing, familiar activity.

3.3.2 The Radiation Protection Imaginary

While artistic creation is always based on the symbolisation of an imaginary relationship to a real object [1], we argue that Castan grasps the imaginary meanings within which the doctrinal argumentation and commitment of SPR agents makes sense [20].

These illustrations are designed to go beyond safety messages and instil the beliefs that guided the activity of the SPR, by representing an ambiguous relationship to violence. Indeed, the danger of radiation or contamination can be both repellent or attractive; the threat must always be kept at bay but can never be eradicated. Such ambiguity can be compared to Girard's definition of the "sacred", namely "everything that controls man, all the more so because man believes he is more capable of controlling it" [8], 51. In this sense, we must not get too close to the sacred, because it unleashes violence, on the other hand, we must also not distance ourselves too far, because it is the foundation of institutions that protect against violence [21].

Radiation protection doctrine was founded on the idea of perfect control, which can be seen in the SPR attempts to legitimise its work. The objective of educating staff and the public was to eradicate the psychosis generated by nuclear power, in order to "show how man, who has succeeded in liberating new forces, can also protect himself effectively" [15]. However, these forces, which are omnipresent at Marcoule, began to manifest insidious effects and proved to be at the limit of what can be measured. The SPR began to see its work as a fight against a formidable enemy that was difficult to represent. Developing a radiation protection doctrine therefore went beyond a question of engineering or processing measurements. Rather, it required developing a collective understanding of a set of representations of its work with an imperceptible, terrifying object. The question was how to maintain a belief in the power of human beings and technical objects that acted as intermediaries in a violent relation with a natural phenomenon. Through the analysis of a series of posters, the following section illustrates these imaginary meanings and this ambiguous relationship to radioactive risk.

3.4 The Representation of Personal Radiation Measurement Equipment

We chose to study the SPR's posters because they were seen as the "most effective direct means of action in the field of information" [10]. For the service, they raised the awareness of workers regarding the specific nature of radioactive risks: "Experience shows that safety has benefited greatly" [16].

The poster is mostly viewed the first time; the second, it is already just a vague reminder, and "even then, if its presence does act on the subconscious, it will have faded into the background" [15]. To compensate for the posters' short-term impact,

Fig. 3.1 *Don't forget your guardian angels* (1959). VRH 2014-04-009. CEA/J. CASTAN

the SPR ensured that they were constantly renewed and defined criteria to increase their effectiveness. Specifically, to ensure that the message would be remembered, the text had to be short and linked to a single topic.

Here, we focus on a series of posters on the topic of dosimeter films and pens (Figs. 3.1, 3.2, 3.3, 3.4, 3.5 and 3.6).[3] Films are passive dosimeters. Once developed, the degree of blackening indicates the irradiation dose received by the agent. Pens are a pocket, electronic dosimeter. They provide workers with an immediate reading of the amount of radiation received. At the individual level, these detectors are fundamental because they determine the pace of work and operational constraints designed to limit the contamination or irradiation of agents.

The posters contain one or more conjugated verbs (except for Fig. 3.5). The affirmative form predominates, and texts are addressed directly to the recipient (use of the second person plural). The text and background are different colours to facilitate reading. Castan uses punctuation, underlines important words or modifies the case of a term. He prefers to only use a few colours and contrasts in order to attract the eye. Posters always contain at least one character, who is either a human drawn in whole or in part, or an anthropomorphic dosimeter. The background is usually either plain or composed of geometric shapes, sometimes supplemented by decorative elements or objects with no connection to a nuclear power plant.

---

[3] The series includes eight posters. In this publication, we present six, reproduced with the permission of the CEA. The two posters not reproduced are entitled *Remember, the danger is invisible* (1962) and *Is your film wet, torn, heated, etc. Tell the SPRAR* (1967).

Fig. 3.2 *Your pen is your friend.* Check it out! (1961). VRH 2014-04-021. CEA/J. CASTAN

Fig. 3.3 *If you lose a film or pen notify the SPR* (1961). VRH 2014-04-013. CEA/J. CASTAN

Fig. 3.4 *Hand over your films* (1961). VRH 2014-04-101. CEA/J. CASTAN

Fig. 3.5 *Your sixth sense!* (1962). VRH 2014-04-35. CEA/J. CASTAN

Fig. 3.6 *Don't let me fall, my heart is weak!* (1964). VRH 2014-04-69. CEA/J. CASTAN

Castan avoids harsh imagery and adopts a humorous approach. He combines amusing and unusual images, drawing his inspiration from religion, fairy tales, medieval stories, esotericism, science fiction, pop culture or the history of France. More specifically, we can identify three main iconographic elements within the series: anxiety, anthropomorphism and sublimation.

3.4.1 Anxiety

According to the centre's director: "[radioactive] *risk is everywhere in this hunt where the presence of perfidiously camouflaged game can only be perceived by the hunter through his electronic 'hunting dog'*" [6]. But how could the service monitor the health of workers if they did not use or hand in their dosimeter films? This concern appears in the service's reports that note, in December 1958, that 10% of films were not handed in.[4] In the following months, this rate decreases. In June 1960, only 0.45% of films were lost and 0.29% were not returned.[5] In subsequent years, reports confirm that good practices became established.

---

[4] Rapport d'activité, 2 janvier 1959, VRH 2009-043-175.
[5] Rapport d'activité, 1er juillet 1960, VRH 2009-043-193.

The SPR attributed much of this success to the effectiveness of its education programme and poster campaigns. Consequently, here we examine how Castan represented the anxiety induced by the loss of a dosimeter. In Fig. 3.3, the detectors, shown as childish silhouettes, are desperate. Like Hansel and Gretel, they have been left to their fate in a hostile environment, embodied in the rigidity of the composition and its background colours. The title symbolises a staircase to the afterlife, which the dosimeters, equipped with blue wings, are about to enter.

This interpretation raises the question of the role of death in Castan's work. Although he never explicitly mentions it, he suggests it, for example by the repeated use of black backgrounds (Figs. 3.1 and 3.6). The representation of death sometimes depends on a subtle detail, as in Fig. 3.4. The name on the dosimeter indicates "Vercingetorix", a French national hero. Workers are thus invited to return their films, as the Gallic leader handed over his weapons to the Roman conqueror. However, this historical reference is disturbing, as the character's surrender did not save him from death. The image is all the more disturbing as the film leads the agent to lose his cover and "expose" himself. Although the poster asks the recipient to comply with the SPR's requirements, it also seems to implicitly question the effectiveness of dosimeters and dosimetry, whose accuracy continued to be improved during the 1960s.

3.4.2 Anthropomorphism

Anthropomorphism creates an analogical relationship between the human and their dosimeter. In Fig. 3.1, the difference between the two angelic silhouettes is reminiscent of Stan Laurel and Oliver Hardy. The reference to the comic duo introduces a lighter note and reinforces the idea that the two figures are inseparable.

Dosimeters are also presented as a dependable colleague. A friendly relationship can be established between these human and non-human partners (Fig. 3.2), or even a romantic alliance (Fig. 3.6). Castan draws upon the story of Rapunzel, trapped in a tower, which is climbed by the pen. But unlike the princess in the Grimm brothers' fairy tale, she remains unmoved by the advances of her suitor, although he promises to save her. Nevertheless, the promise of love remains, indicated by Rapunzel's red dress, which is the same colour as the precious gift offered by the dosimeter.

Anthropomorphism is thus a means of emphasising the usefulness of dosimeters, non-human collaborators and guardian angels of workers. However, the angel illustrates all the ambiguity of the symbolism of Castan. The angel, symbol of invisible powers, in some cases represents the dosimeter with protective powers, and in others the radiation that must be protected, as in *Remember, the danger is invisible*, a poster drawn in 1962.

3.4.3 Sublimation

Given the relationship to the unrepresentable, the agent must place his faith in technical objects that act as intermediaries between him and an inaccessible world. Castan therefore sublimates dosimeters. In Fig. 3.1, the pen and film are celestial creatures that protect workers' lives. Figure 3.2 represents the pen as a psychic with a crystal ball. Here, the object (pen) communicates with intangible forces—radiation. Its eyes are closed, suggesting that it is in a trance. Its finger underlines its submissiveness and points to a celestial elsewhere, a higher reality that only it knows. The characters do not seem to belong to the same world. The pen is shown on a yellow background, while the agent is shown on a green background, dividing the space occupied by each character. A green owl is shown on the yellow background. This symbolises both the clairvoyance of the object and the wisdom of the worker who came to consult his friend. The animal acts as an intermediary between the two worlds, while the yellow teapot on the green background suggests the convivial relationship between human and non-human actors.

The ability of dosimeters to save lives is also presented. The spacing of Gainsbourg's[6] hands, in Fig. 3.5, evokes religious iconography. The pen, by absorbing radiation, is sanctified, suggested by its glowing halo. The smoking singer seems to be thanking the device that provides security—reinforced by the brightness of the blue background—to those who remember to use it.

In a more warlike register, the dosimeter pen becomes, in *Remember, the danger is invisible*, the only "weapon" available to the operator to protect himself. Unable to see the winged imps who come to taunt him, he has to place his faith in his equipment. In general, the posters therefore associate the dosimeters with border objects, thus materialising the mystery that connects the visible to the invisible.

3.5 Conclusion

The iconography used in Castan's posters reflects an approach to risk prevention that is consistent with the sensitivities of the time. At the turn of the 1950s, poster artists abandoned brutal images in favour of other approaches, such as showing ways to save your own life or humour [2].

The approach has raised questions about the underlying rationale. The entertaining dimension of Castan's posters exposes the SPR's perception of their colleagues: "*SPR officers thought that workers could not be trusted to take safety issues seriously*" [11], 196. Do Castan's posters therefore seek to infantilise CEA workers? We argue that they do not. On the one hand, the education programme aimed to empower staff, on the other hand, posters captured the radiation protection imaginary and reflected an ambiguous relationship with radioactive risk. The presence of this ambiguity supposes that the viewer was able to perceive the multiple levels

---

[6] Serge Gainsbourg (1928–1991) was a French singer, pianist, composer, poet, painter and actor.

of meaning in the image—why else represent it? This observation suggests that the thesis of infantilisation should be refuted.

This ambiguity does not reflect a form of dissent from Castan. Rather, it seems to reflect a degree of collusion between the poster's designer and the worker who viewed it. For example, Fig. 3.5 implicitly states that the SPR knows that staff smoke in the working area. However, while this disregard for regulations appears to be tolerated, the service is uncompromising with respect to the use of the dosimeter.

Given the lack of data regarding the effectiveness of posters, a content analysis can shed light on the procedures used to educate agents. These images, beyond their evocative power and ability to represent an imperceptible risk, reflect, to a greater or lesser extent, the day-to-day working environment. As such, the posters work iconographically because they speak for the pioneers of a booming industry.

References

1. D. Anzieu, *Le corps de l'œuvre* (Gallimard, Paris, 1981)
2. N. Blétry, Ceci n'est pas un risque, in *Cultures du risque au travail et pratiques de prévention.* ed. by C. Omnès, L. Pitti (Presses Universitaires de Rennes, Rennes, 2009), pp.155–172
3. Commissariat à l'Energie Atomique, *Rapport Annuel* (CEA, CEA Saclay, 1957)
4. C. Davillerd, La compréhension d'affiches de sécurité : enquête en industrie. In *Exposition internationale sur l'impact de l'affiche de sécurité* (1986)
5. F. Duhamel, Les problèmes de radioprotection et leur enveloppe. Bulletin d'Informations Scientifiques et Techniques (BIST) **39**, 2–10 (1960)
6. M. Gervais de Rouville, Editorial. BIST 72–73: 1–4 (1963)
7. M. Gervais de Rouville, Le centre de production de plutonium de Marcoule. Energie Nucléaire **5** (1964)
8. R. Girard, *La violence et le sacré* (Fayard, Paris, 2010)
9. S.J. Guastello, Do we really know how well our occupational accident prevention programs work? Saf. Sci. **16**, 445–463 (1993)
10. C. Guérin, *Formation et éducation du personnel en matière de radioprotection sur le centre de Marcoule* (CEA, CEA Marcoule, 1964)
11. G. Hecht, *Le rayonnement de la France, énergie nucléaire et identité nationale après la Seconde Guerre mondiale* (La Découverte, Paris, 2004)
12. F. Lamare, F. Guarnieri, A. Portelli, Normaliser la protection des travailleurs du nucléaire dans les années soixante: le premier bréviaire de radioprotection. Revue Générale Nucléaire **4**, 58–59 (2018)
13. A. Portelli, F. Guarnieri, Quand le SPR de Marcoule racontait le nucléaire en bande dessinée. Revue Générale Nucléaire **1**, 72–77 (2015)
14. A. Portelli, S. Travadel, F. Guarnieri, C. Parizel, Peindre la radioprotection à l'aube du nucléaire. Revue Générale Nucléaire **4**, 56–61 (2017)
15. J. Rodier, J. Castan, C. Guérin, Information et éducation en matière de radioprotection. BIST **72–73**, 91–98 (1963)
16. J. Rodier, J.-P. Chassany, H. Peyresblanques, R. Estournel, R. Court, *Expérience en matière de radioprotection tirée de 8 ans de fonctionnement d'un centre de production du plutonium* (CEA, CEA Marcoule, 1966)
17. K.L. Saarela, A poster campaign for improving safety on shipyard scaffolds. J. Saf. Res. **20**, 177–185 (1989)
18. P. Saint Raymond, Une longue marche vers l'indépendance et la transparence, l'histoire de l'Autorité de sûreté nucléaire française. La Documentation Française, Paris (2012)

19. R.G. Sell, What does safety propaganda do for safety? A review. Appl. Ergon. **8**, 203–214 (1977)
20. S. Travadel, C. Parizel, A. Portelli, F. Guarnieri, Doctrine de la radioprotection à l'aube de l'industrie nucléaire : récit en images. Cahiers de Narratologie **32** (2017). https://doi.org/10.4000/narratologie.7772
21. S. Travadel, A. Portelli, C. Parizel, F. Guarnieri, Les figures de l'infime: la radioprotection en images. Techniques et culture **68**, 111–129 (2017)

Open Access This chapter is licensed under the terms of the Creative Commons Attribution 4.0 International License (http://creativecommons.org/licenses/by/4.0/), which permits use, sharing, adaptation, distribution and reproduction in any medium or format, as long as you give appropriate credit to the original author(s) and the source, provide a link to the Creative Commons license and indicate if changes were made.

The images or other third party material in this chapter are included in the chapter's Creative Commons license, unless indicated otherwise in a credit line to the material. If material is not included in the chapter's Creative Commons license and your intended use is not permitted by statutory regulation or exceeds the permitted use, you will need to obtain permission directly from the copyright holder.

Chapter 4
Ways of Seeing (and Not Seeing) Safety

Patrick Waterson

Abstract This chapter seeks to offer some explanation for the ubiquity of different types of visual representations in safety science. In particular, the chapter focuses on what these tell us about the thinking of safety researchers and practitioners, as well as how diagrams and other visual material influence their use of safety methods and tools.

Keywords Visual representations · Safety models and methods · Accident analysis

4.1 Introduction

In his book 'Man-Made Disasters', the safety theorist and organisational sociologist Barry Turner remarked that *'a way of seeing … is always also a way of not seeing'* (Baddeley and Hitch [1], p. 49). Turner was talking about what he termed the 'decoy problem'—the fact that attention may be paid to a well-defined problem or source of danger, but this also distracts from other more potentially dangerous problems lurking in the background. In this chapter, I want to focus on a slightly different set of issues and questions centred around 'seeing' in safety, in particular the use of visual representations in safety tools, models and methods. In particular, the chapter focuses on two main questions:

1. What does the use of diagrams and other forms of visual representations by safety researchers tell us about their thinking and the underlying theory of safety they seek to promote?
2. How do the various types of visual representations used in safety models, methods and tools work influence their use by researchers and practitioners?

P. Waterson (✉)
Human Factors and Complex Systems Group, Loughborough University, Loughborough, UK
e-mail: P.Waterson@lboro.ac.uk

© The Author(s) 2023
J.-C. Le Coze and T. Reiman (eds.), *Visualising Safety, an Exploration*,
SpringerBriefs in Safety Management, https://doi.org/10.1007/978-3-031-33786-4_4

4.2   John Berger and 'Ways of Seeing'

There is a well-established tradition within the History of Art which focuses on the influence of pictorial and other forms of visual representations on our ability to perceive, understand and interpret works of art. The art historian Ernst Gombrich for example, in his book 'Art and Illusion' (1960) argued for the importance of 'cognitive schemata' [2] in analysing works of art. Gombrich claimed that artists learn to represent the external world by learning from previous artists, and as a result representation is often achieved using stereotyped figures and methods.

More recently, the artist Bridget Riley has argued that the use of colour and black and white in her work has the power to trigger perceptual and cognitive illusions and other visual stimulations in the viewer [3]. One of the most important works of the writer and cultural theorist John Berger (1926–2017) was a book based on a television series which was screened in the UK in the early 1970s. 'Ways of Seeing', Berger [4] set out to criticise traditional Western cultural aesthetics and raised questions about hidden ideologies in visual images (e.g., magazine advertisements). In one episode of the programme, Berger showed the continuities between post-Renaissance European paintings of women and modern-day posters and advertisements, by juxtaposing the different images and showing how they similarly rendered women as objects. The book was partly written as a riposte to the more traditionalist view of the Western artistic and cultural canon (e.g., the work of Kenneth Clark and his book/TV series 'Civilisation', 1969) and the TV programmes and book criticise traditional Western cultural aesthetics by raising questions about hidden ideologies in visual images. Berger offered a Marxist alternative which shifted attention towards the cultural messages and sub-meanings which are embodied when we look at objects and art.

4.3   Visual Representations in Accident and Safety Research

Berger's analysis of the process of seeing art and revealing some of the implicit hidden meanings in paintings and other forms of art, might serve as a useful basis with which to probe deeper into the theoretical roots and origins of many of the types of visual representations used in the world of safety. In a series of articles describing the history of safety science, Swuste et al. [5, 6] make use of a wide variety of visual materials including posters, warning signs and cartoons to illustrate the way in which theoretical and methodological approach to understanding accidents has changed over the last century. Similarly, [7] argues that safety science and occupational safety and health relies heavily on visualisation as a means to communicate safety messages and sometimes act as metaphors and boundary objects. Warning signs, for example, rely heavily on icons, pictorials and other visual materials and the models, methods and tools used by safety practitioners. It is also a testament to their dominance that some models (e.g., the Bird/Heinrich triangle) still remain popular even in the face of significant criticism within the scientific community (e.g., [8]).

4.4   Why Are Visual Representations So Popular?

One of the most obvious reasons why visual representations feature so prominently in safety can be attributed to the Chinese proverb 'a picture is worth 10,000 words'. Because we think with the help of images, pictures, diagrams and other forms of imagery help us to reflect deeply beyond words. Visual representations may also be augmented and act as 'cognitive aids' through the use of metaphors [9], for example Bow-Tie models or Reason's three bucket model of human error. When they are solving problems, human beings use both internal representations, stored in their brains, and external representations, recorded on a paper, on a blackboard, or on some other medium [10, 11]. Amongst engineers, scientists and designers visual representations are also commonly used as a means of structuring and communicating complex problems [12–14]. Visual representations serve as 'boundary objects' which are open to interpretation across the various communities and specialisms (e.g., occupational safety and health managers, risk managers) involved in safety, but also serve as a common focal point supporting cross-disciplinary communication and collaboration [7, 15]. Henderson [16] further developed the concept of 'meta-indexicality' in order to underline the ability of visual representations to combine many diverse levels of knowledge and to serve as a meeting ground for many types of disciplines and individuals.

4.5   The Evolution of Two Safety Models (Swiss Cheese Model and Accimaps)

Probably the most well-known model in safety is James Reason's Swiss Cheese Model. Reason developed at least two different versions of the model over the course of the 1980s and 1990s. What is interesting about the earlier version of the SCM is that it retains an element of the types of box models of cognition common in psychology in the 1960s and 1970s (e.g., [17] model of working memory). Similarly, the box model or flow chart is also reminiscent of the various types of error taxonomies which were developed in the 1980s (e.g., Reason's 1988 GEMS model, Rasmussen's 1983 SRK framework), as well as the fault trees commonly used to assess risk in the nuclear and other high-risk industries. By the time we reach the late 1990s, the model has become a metaphor which supports a more systems-oriented way of thinking about latent and active pathways ('holes') across the various levels within the system ('slices') and the role they play in causing human error.

A final example of how safety models shift and evolve over time and what this tells us about implicit theories of safety is [18] Accimap model. An earlier version shown in [19] was a hand-drawn model of the abstraction hierarchy, elements of which formed the basis of Cognitive Work Analysis, Cognitive Systems Engineering and the Risk Management Framework (Rasmussen [18]). The RMF also served as the basis with which to develop the first Accimaps, a method for analysing accidents which has proved popular and spawned many variations and 'remixes' [19].

4.6   Augmenting and Extending How We Use and Evaluate Visual Representations in Safety

Many of the visual representations mentioned in this chapter are used as methods for analysing and sometimes investigating accidents. One of the most common ways of facilitating comparison and evaluation between the methods is to look at their scientific properties (e.g., validity, reliability, coverage of systems thinking components). This proves often to be problematic [20] and from the point of view of safety practitioners, often misses the point. Safety investigators and risk assessors, for example, often cite the usability of the method as more important than scientific concerns. Being able to learn the method quickly and how resource-intensive, it is are also important considerations. One way forward then may be to shift the focus of evaluations of safety models and methods away from strict scientific criteria and more towards ease of use and usability in general. Green [21], for example, describes a set of cognitive dimensions of notations which are designed to provide a lightweight approach to analyse the quality of a design, rather than an in-depth, detailed description. They provide a common vocabulary for discussing many factors in visual notation, user interfaces or the design of programming languages. The dimension 'role expressiveness', for example, is defined as 'how obvious is the role of each component of the notation in the solution as a whole?' and an associated question is 'When reading the notation, is it easy to tell what each part is for?' Are there some parts that are particularly difficult to interpret? Likewise, the 'viscosity' of a notation might be assessed through questions such as 'how readily can required parts of the notation be identified, accessed and made visible?'. Finally, the 'abstraction gradient' dimension [22] might be applied in order to assess the degree to which methods or models allow further refinement and elaborations ('abstractions') of causal components. The sort of trade-offs involved in different methods as might take the form of comparing for example, FRAM [23] (possibly high on 'role expressiveness', possibly low on 'viscosity') with other methods (e.g., Accimap—possibly scoring high on both dimensions). In other words, these types of what might be called 'low fidelity' criteria might prove helpful in further improving the visual representations used in safety models, as well as assessing their fitness for purpose within the practitioner community. It might also be one way in which we break out of what is sometimes seen as rather dry and sterile debates surrounding which method is better for accident analysis as compared to another [24].

4.7 Some Conclusions: Ways of Seeing Safety

The diversity of visual representations which are used in safety science is evidence enough that we have evolved multiple ways of 'seeing safety' in the sense that John Berger suggested. Models and methods provide us with one of the many lenses with which we can look at safety and afford a way of what Barry [1] described as *'turning the kaleidoscope'*. They also raise a number of challenges for the future, not least how to deal with ever-increasing complexity in terms of systems, technology and organisations. How we trade off complexity against simplicity in our attempts to model safety in these systems will continue to be an important question for the foreseeable future. Some might argue that some of our present models and methods (e.g., FRAM, STAMP) have gone too far and widened an already large gap between safety researchers and practitioners [25], whilst at the same time casting the net too far in terms of how we might design solutions which might prevent accidents or attempt to link organisational factors to casual factors contributing to accidents [26, 27]. How we narrow some of these gaps, as well as wider questions about how visual representations structure and shape our views on safety will continue to occupy our attention in the coming years.

References

1. B.A. Turner, *Man-Made Disasters* (Wykeham Publications, London, 1978)
2. F.C. Bartlett, *Remembering: A Study in Experimental and Social Psychology* (Cambridge University Press, Cambridge, 1932)
3. B. Riley, E.H. Gombrich, The use of colour and its effect: the how and the why. Burling. Mag. **136**(1096), 427–429 (1994)
4. J. Berger, *Ways of Seeing* (Penguin Books, Harmondsworth, 1972)
5. P. Swuste, C. van Gulijk, W. Zwaard, Safety metaphors and theories, a review of the occupational safety literature of the US, UK and The Netherlands, till the first part of the 20th century. Saf. Sci. **48**, 1000–1018 (2010)
6. P. Swuste, C. van Gulijk, W. Zwaard, Y. Oostendoorp, Occupational safety theories, models and metaphors in the three decades since World War II, in the United States, Britain and the Netherlands: a literature review. Saf. Sci. **62**, 16–27 (2014)
7. J.-C. Le Coze, Visualising safety, in *Safety Science Research: Evolution, Challenges and New Directions*, ed. by J.-C. Le Coze (CRC Press, Taylor and Francis, Boca Raton, 2019)
8. A. Hale, Conditions of occurrence of major and minor accidents: urban myths, deviations and accident scenario's. Tijdschrift voor toegepaste Arbowetenschap **15**, 34–41 (2002)
9. G. Lakoff, M. Johnson, *Metaphors We Live By* (Chicago University Press, Chicago, 1980)
10. J. Larkin, H.A. Simon, Why a diagram is (sometimes) worth ten thousand words. Cognit. Sci. **11**, 65–99 (1987)
11. Z.W. Pylyshyn, What the mind's eye tells the mind's brain: a critique of mental imagery. Psychol. Bull. **80**, 1–24 (1973)
12. K. Baynes, F. Pugh, The art of the engineer. Accompanying Booklet to a Welsh Arts Council Touring Exhibition (1978)
13. L. Bucciarelli, Between thought and object in engineering design. Des. Stud. **23**, 219–231 (2002)

14. M. Kemp, Seeing and picturing: visual representation in twentieth-century science, in *Science in the Twentieth Century*, eds. by J. Krige, D. Pestre (Routledge, London, 2014)
15. S.L. Star, J.R. Griesemer, Ecology, 'translations' and boundary objects: amateurs and professionals in Berkeley's museum of vertebrate zoology. Soc. Stud. Sci. **19**, 387–420 (1989)
16. K. Henderson, *On Line and On Paper: Visual Representations, Visual Culture and Computer Graphics in Design Engineering* (MIT Press, Cambridge, 1999)
17. A.D. Baddeley, G. Hitch, Working memory, in *The Psychology of Learning and Motivation: Advances in Research and Theory*, vol. 8, ed. by G.H. Bower (Academic Press, New York, 1974), pp.47–89
18. J. Rasmussen, Risk management in a dynamic society: a modelling problem. Saf. Sci. **27**, 183–213 (1997)
19. P.E. Waterson, D.P. Jenkins, P.M. Salmon, P. Underwood, 'Remixing Rasmussen': the evolution of Accimaps within systemic accident analysis. Appl. Ergon. **59**, 483–503 (2017)
20. A.P.G. Filho, G.T. Jun, P.E. Waterson, Four studies, two methods, one accident: an examination of the reliability and validity of Accimap and STAMP for accident analysis. Saf. Sci. **113**, 310–317 (2019)
21. T.R.G. Green, Cognitive dimensions of notations, in *People and Computers V*, eds. by A. Sutcliffe, L. Macaulay (Cambridge University Press, Cambridge, 1989)
22. T.R.G. Green, M. Petre, Usability analysis of visual programming environments: a 'cognitive dimensions' framework. J. Vis. Lang. Comput. **7**(2), 131–174 (1996)
23. E. Hollnagel, *FRAM: The Functional Resonance Analysis Method: Modelling Complex Socio-Technical Systems* (Ashgate, Farnham, 2012)
24. P.E. Waterson, C.W. Clegg, M. Robinson, Trade-offs between reliability, validity and utility in the development of human factors methods, in *Human Factors in Organizational Design and Management XI*, eds. by O. Broberg, N. Fallentin, P. Hasle, P.L. Jensen, A. Kabel, M.E. Larsen, T. Weller (IEA Press, Santa Monica, 2014)
25. P.E. Waterson, Bridging the gap between research, policy and practice in health and safety. Policy Pract. Health Saf. **14**, 97–98 (2016)
26. S. French, I. Steel, Looking beyond the obvious: the investigation of organisational factors following an accident, in *Paper Presented at the International Railway Safety Conference, Dublin* (2018)
27. J.T. Reason, Are we casting the net too widely in our search for the factors contributing to errors and accidents? In *Nuclear Safety: An Ergonomics Perspective*, eds. by J. Misumi, B. Wilpert, R. Miller (CRC Press, Boca Raton, 1999)

Open Access This chapter is licensed under the terms of the Creative Commons Attribution 4.0 International License (http://creativecommons.org/licenses/by/4.0/), which permits use, sharing, adaptation, distribution and reproduction in any medium or format, as long as you give appropriate credit to the original author(s) and the source, provide a link to the Creative Commons license and indicate if changes were made.

The images or other third party material in this chapter are included in the chapter's Creative Commons license, unless indicated otherwise in a credit line to the material. If material is not included in the chapter's Creative Commons license and your intended use is not permitted by statutory regulation or exceeds the permitted use, you will need to obtain permission directly from the copyright holder.

Chapter 5
Representations, Metaphors and Slogans: From Organisational Safety to Societal Resilience

Torgeir Kolstø Haavik

Abstract We discuss what it is with representations of safety that makes them so powerful, and what is at stake when representations travel across contexts and scales. The discussion uses the sharp end/blunt end metaphor as a central case.

Keywords Visualisation · Representation · Metaphors · Immutable mobiles

5.1 Visualisation and Representation

Visualisation is an important ingredient in science, as in most creative activities. Basically, creativity requires *imagination*, and visualisations can be extremely powerful. Think, for example, of the recently published, first ever, photography of a black hole [1] in the universe. Undoubtedly, this representation of a physical phenomenon—whose existence has previously been evidenced mainly by theory—will be of great importance not only for further scientific work on the relativity theory, but also for raising money for space research.

Visualisation is a linguistic contraction of the terms *visual* and *representation*. What the contraction elegantly hide is that visualisation has both surface and depth. The visual aspect is the surface, while representation goes deeper. Think for example of the famous Munch work *The Scream*; while the visual aspects of that painting are surely interesting—the technique, the colours, the proportions—the most intriguing issue lies beneath the surface; how shall we understand The Scream, what kind of psychological condition does Munch portray, what occasioned the work, and how can we ascribe personal and collective relevance to the painting? What does this two-dimensional art work—that has turned into an obligatory passage point in art science and art history—*represent*?

T. K. Haavik (✉)
NTNU, Trondheim, Norway
e-mail: torgeir.haavik@samforsk.no

© The Author(s) 2023
J.-C. Le Coze and T. Reiman (eds.), *Visualising Safety, an Exploration*,
SpringerBriefs in Safety Management, https://doi.org/10.1007/978-3-031-33786-4_5

In this chapter, I will focus on representation. I will discuss what it is with representations of safety that makes them so powerful, and what is at stake when representations travel across contexts and scales. A widely known representation—or metaphor—in safety science, the sharp end/blunt end metaphor, will serve as a case.

Representations have received much attention in science studies. A brief review of representations and their potential as *immutable mobiles* in scientific practice can therefore be worthwhile, before proceeding with the discussion of representations in safety science.

5.2   Representations as Immutable Mobiles

The books *Representation in Scientific Practice* [2] and *Representation in scientific practice revisited* [3] are a good place to start to understand the role of representations in science. In the preface of the latter, Lynch and Woolgar reflect upon the activities and the discourse on representations among the contributors leading up to the book. There were different suggestions with regard to the book's title, but.

> (…) we decided to focus the issue on *representation* in scientific practice. Inspired in part by the growing interest in visualization, we also wanted to bring into play close studies of verbal interaction at the lab bench (or field site), as well as analyses of the literacy and pragmatic relations among texts, depictions, and activities. [3]

The citation reminds us of the close relationship between visualisations, texts and actions, in the sense that they are all representations.

Central aspects of representations are perception, suggestiveness and communication—the way visualisations describe and explain phenomena—and these aspects have received a significant proportion of the attention in the treatment of visualisation in safety research, e.g., by Le Coze [4, 5]. Less attention has been given to the *circulation* of visualisations in safety science.

In *Visualisation and cognition: thinking with eyes and hands*, Latour [6] explores the *circulation* of such types of representations—frequently labelled inscriptions: drawings, diagrams, plots, images, maps, signs, photographs—in science. Latour underscores that the characteristics of inscriptions and ways of perceiving them in this context is not a question of perception, but of "something deeper" [6]: particularly it is about mobilisation and stability. Inscriptions in science may be highly *mobile*: science is about capturing essential aspects, characteristics, connections and causalities of the empirical world, to transform them into theoretical formulations that may easily circulate across a larger scientific community. Inscriptions may be *immutable*: immutable inscriptions are such that their shapes and contents are left unchanged as they circulate in the scientific community. In addition, inscriptions may be scalable: The *scale* of inscriptions may be *modified* without any change in internal proportions.

Latour labels scientific inscriptions that are both mobile and immutable *immutable mobiles*, and he ascribes to them an enormous significance in science and research:

> It seems to me that most scholars who have worked on the relations between inscription procedures and cognition, have, in fact, in their various ways, been writing about the history of these immutable mobiles. [6]

Together, such aspects of inscriptions—and all representations that can count as immutable mobiles—have played an important role in the development and communication of modern science, for example in geography (maps—e.g., Ptolemy's 2nd-century world map [7]), in chemistry (formulas, Mendeleev's [8] periodic table) and in sociology (e.g., four-field tables).

However, the ingenuity of immutable mobiles comes at a price. By ensuring mobility and immutability of a phenomenon, by turning an unruly three-dimensional world into a two-dimensional, stable representation that can travel without much friction, enter into and stabilise wider webs of knowledge, one runs the risk of inscribing ontology in ways that produce challenges when the representation arrives in a different context. This challenge will be discussed in the remainder of this chapter.

5.3 Representation in Safety Science: The Sharp End/Blunt End Metaphor

As Le Coze [4] shows, representations have had a tremendous impact on the developments and discourses in safety science, and some of the most influential representations function almost as obligatory passage points; think for example of Perrow's [9] risk matrix and Reason's [10] Swiss cheese model.

In the following, we shall try to extend the perspective on representations in safety science, and reflect on their strengths and weaknesses as scientific objects, by viewing them as immutable mobiles. As a case, we will use the sharp end/blunt end representation—or metaphor—introduced in safety science by James Reason and later used by many others [11–14].

While the sharp end/blunt end metaphor does not have one particular canonical visualisation associated with it (but see Fig. 5.1 below), it is so visual in its expression—and by now so embodied—that it immediately evokes a canvas onto which almost any safety scientist or safety practitioner can envisage the scenery of the sharp end/blunt end.

The sharp end/blunt end metaphor assumes both a hierarchical and a linear view on sociotechnical systems. The sharp end refers to the context where work is carried out and where the consequences of actions manifest themselves directly and immediately—*here and now*. The blunt end, on the other hand, invokes spheres of the organisation and beyond that do not directly take part in work at the sharp end, but it influences the personnel, equipment and general conditions of work at the sharp end. The blunt end is *there and then* [15] (see Fig. 5.1).

The sharp end/blunt end bifurcation may, by its mere existence, lend support to arguments for different perspectives on safety, from compliance perspectives

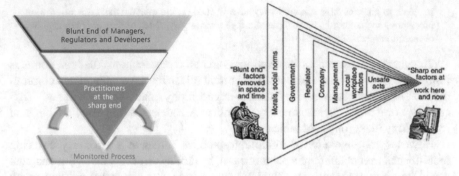

Fig. 5.1 Sharp end/blunt end, from Woods et al. [16] (left) and Hollnagel [14] (right)[1]

that promote the possibility of managing safety from the blunt end, to practice based perspectives that emphasises the role of situated practice, adaptations and adjustments—as mostly accentuated in resilience studies.

The sharp end/blunt end metaphor is highly mobile and combinable. If we consider Rasmussen's [17] famous hierarchical representation of a sociotechnical system, the sharp end/blunt end metaphor has been tightly integrated into it—or vice versa; not only as a way of drawing, but as a way of thinking in linear and hierarchical terms. So incorporated is the sharp end/blunt end metaphor in our thinking that it is close to an obligatory passage point for contextualising risk and safety in a landscape of organisational structures, regulations and practice.

The sharp end/blunt end metaphor is one of those representations that are difficult to bypass, although one does not necessarily subscribe to its foundational linear/hierarchical ontology. This ontology, draped in controversies, does not seem to stand in the way for using the metaphor also in contexts where the sociotechnical arrangements and dependencies are thought of as *systemic*. As such, and for the following reasons, the sharp end/blunt end visualisation is very powerful;

- It is highly mobile: One need not say more than "sharp end" to make commensurable series of imaginaries of operators working in such different locations and situations as in control rooms of nuclear power plants, in airline cockpits, in the midst of forests burning or at the deck of aircraft carriers. At the same time, we imagine those blunt end managers far away that have structured the conditions that the operators work under.
- It has immutable qualities; traversing different contexts, scales and purposes, it is capable of remaining its associative vectors that reproduce imaginaries

[1] Illustration in the left panel is used with permission from the author. Illustration in the right panel is reproduced from Barriers and Accident Prevention, 1st Edition, published by Routledge. © Erik Hollnagel, [14]. Reproduced by arrangement with Taylor & Francis Books UK. All rights reserved. Both illustrations are excluded from our open access license.

of linear/hierarchical causality even in circumstances where system functionality and descriptions are explicitly otherwise—think for example of any FRAM visualisation (e.g., [18, 19]).

- It is scalable; the sharp end and the blunt end are relative terms that apply to settings that are highly different in scale, but still internally consistent. As noted by Karlene Roberts, "Everybody's blunt end is somebody else's sharp end" [20]. Hence, if the blunt end in one setting is the administrative level at a hospital, the sharp end may refer to the work of paramedics at an accident site. In relation to the blunt end of the World Health Organisation's headquarters in Geneva, however, the local hospital can naturally be considered as the sharp end. In this way, the sharp end/blunt end metaphor can easily be circulated across contexts with limited need to undertake comprehensive work to adapt scales between the contexts. As we shall see, however, there is a challenge associated with this transportation across contexts and scales.

5.4 The Twist of the Sharp End/Blunt End Metaphor

Intuitive as the sharp end/blunt end metaphor may seem, it is more ambiguous than comes into expression in the daily use of it. Applied to a micro scale and a mesoscale—that is—up to a level of organisational life—and in the context of organisational language—ambiguity is not particularly conspicuous.[2] Those practitioners portrayed in the sharp end in Fig. 5.1 tend to be those who "actually interact with the hazardous process in their roles as pilots, physicians, spacecraft controllers, or power plant operators" [12], and those who are in a position to make the necessary adaptations. In an organisational setting, thus, there will often be a straightforward convergence between the intuitive interpretation of the sharp end/blunt end metaphor, and the actual operationalisation of organisational charts of everyday work on the other.

However, as soon as we depart from the realm of organisations and direct the attention towards complex sociotechnical systems, or as we scale up further and transcend the boundaries of organisations and traditional sociotechnical systems and enter into a landscape of societies and global risks, limitations of the metaphor begin to appear. As we, in this new context, review our inventory of representations to make sure that the relationship between *representations* and the *represented* are adequate, we may find that it is no longer obvious what constitutes the sharp end and the blunt end.

One of the most serious challenges we are facing in terms of societal resilience today is the climate change that threatens to seriously change living conditions on earth. One of the hot controversies in that connection is whether the actor-network in the best position to stagger or reverse climate change are that of international

---

[2] Indeed, the idea of complex organisations really working in a linear manner is also mostly abandoned in the safety literature on complex sociotechnical systems, but that does not prevent the sharp end/blunt end metaphor from still being a central reference [15, 21–24].

politics or of "ordinary citizens". If we think of the societal/global system in terms of a sociotechnical system with a sharp end and a blunt end, there immediately seems to be parallels between sharp end operators and citizens, on one hand, and between blunt end managers and top-level politicians on the other. With that parallel, at first it seemingly makes sense to think of non-sustainable citizen behaviour (unsustainable consumerism) as an issue relating to the sharp end, and of lack of regulatory measures as a problem located at the blunt end, among politicians and other global decision makers. The potential of resilience lying in adaptation inclines us to address in particular the sharp end, which would in that case be citizen behaviour. But how well does that framing fit the nature of climate change and the ecosocial systems it takes place within?

The issue here is not whether the climate crisis should be addressed at the sharp end or the blunt end, but rather *what is the sharp end* and *what is the blunt end* of the climate system? Contrary to our (my) hasty assumption above, I shall contend that in the context of climate change—if I am forced to relate to it in terms of sharp and blunt ends—the realm of international politics and decision makers constitute the sharp end, and the realm of the citizens constitute the blunt end.

The sharp end/blunt end metaphor is created with local risks in mind. It is for that reason the traditional "local operator" is the one who has been associated with the sharp end. However, when the risk is global, we need to look for global operators. In the blunt end, we will find those who give the operators mandates—or orders—to act on their behalf, and those are ordinary citizens.

In this perspective, an example of the climate system's sharp end may be constellations such as the United Nations Climate Change Conferences,[3] yearly conferences (Conference of the parties—COP) that assess progress in dealing with climate change. These constellations can be described as the sharp end since.

- They have been arenas for negotiating the Kyoto Protocol and the Paris Agreement—among the sharpest measures so far for dealing with climate change at a global scale.
- *Climate* is a global phenomenon that exists and can only be measured and addressed at a global scale.[4]
- The main tool to analyse global climate change is *The Intergovernmental Panel on Climate Change* (IPCC), whose assessment reports are key scientific inputs into the UN Framework Convention on Climate Change (UNFCCC). The reports are compilations of worldwide climate research from a multitude of disciplines and are thus a holistic evaluation of the climate status at a global scale.

---

[3] These climate change conferences serve as the formal meetings of the parties, the governing body of the UN Framework Convention on Climate Change, just as other conferences of the parties (COP) serve as formal meetings for other international conventions related to societal resilience, such as those regarding desertification, corruption, biological diversity, chemical weapons and non-proliferation of nuclear weapons.

[4] Surely, every single sensor measurement is local, but when we speak of global temperature rise, for example, that is a parameter that exists (is calculated) only at the global scale.

- Actions taken by the COP can have direct consequences for the global climate if they are articulated as binding commitments,[5] because they may alter practices at a global scale very quickly—here and now, so to speak.

In the context of global risks like climate change, the blunt represents "the citizens". Citizens are at the blunt end since.

- Citizens are voters that provide their country representatives with the mandates they bring to the COPs to make sharp decisions.
- Trends and changes (adaptations) in the public opinion seldom stabilise and materialise until they are operationalised into (or at least supported by) laws.
- Local citizens have access to the *weather*, but no direct opportunity whether to estimate nor influence climate other than through the "climate operators" at the sharp end.

In the context of societal resilience in times of climate crisis, the sharp end and the blunt end seem to have switched poles; the operators with the capacity to undertake adjustments and adaptations (of laws and regulations) that make a difference here and now are represented by the political elite and other decision makers, while those at the blunt end—distanced in space and time—that have little direct influence, but provide the political elite with their mandate, are represented by ordinary voting and public opinion forming citizens.

5.5   A Programme for Societal Resilience

The resilience perspective is particularly occupied with work as it is actually carried out [25–27], and to study and understand *work-as-done* and the potential of adaptation one needs to pay particular attention to the actor-network in the sharp end. When the subject for resilience is global risks, we must populate the sharp end/blunt end metaphor carefully for it to ensure representability. Thus, when we—in this age where human activity has a dominant influence on the world's ecosystems and climate change—want to pay attention to the sharp end of the *world-as-done*, we need to look to the work, the adaptations and adjustments that take place on the global arena. Practice studies, then, which have been so popular in the field of resilience, will in this context advise us to study political practices at least as thoroughly as we study citizen practices.

It is not only when travelling across scales—such as from the organisational to the societal—that the representability of visualisations and metaphors is challenged.

---

[5] It should be noted that the conferences have been quite hesitant towards binding commitments, and there has been a lack of enforcement mechanisms.

There is a slogan from the early days of environmental politics and activism, encouraging us to "think globally, act locally".[6] There is an interesting connection between this slogan, the organisational sharp end/blunt end metaphor and the neo-liberal motive that is sometimes—unfortunately—ascribed to the resilience ideology. That slogan echoes from a vantage point where one does not oversee the form scale of global risks, where international politics have few tools for global governance, and where citizens are thought of as consumers of goods instead of producers of politics. *That* is neo-liberalism, and it is unrealistic that such a regime in the long run will prove resilient.

The example of climate change illustrates a challenge that may arise when representations travel too far away from their place of origin; a resilient global society requires global scale adaptations, and such initiatives are in the hands of international politics and decision makers at the sharp end of the global sociotechnical system. Without such sharp end adaptations, one would never see binding agreements on global risks issues such as climate change, nuclear weapons, chemical weapons, and threatened biodiversity. In light of that, a representational slogan for resilience in an age where the true scale of risks is global although the feel of them is local and individual, could be "Think locally, act globally".

The paradox of this text is that the author and probably many readers do not think of systems in terms of sharp end and blunt end at all, as little as we want to distribute causes and effects along the same axis. So why discuss the metaphor at all? The answer to that is highly pragmatic: although we don't believe in it, the metaphor continues to work, just like we in the digital era continue to arrange our world into 0 and 1 s although we know that our lifeworlds are (still for a while) much richer than that. We probably still will have to live with the sharp end/blunt end representation also in the future, but in terms of research method it is advisable to always have in mind Latour's [28] slogan: *follow the actors*. That will—within the limitations of research funding—enable us to capture significant work across the networks of actors all the way from the shop floor to the boardrooms, from the citizens to the UN conferences.

References

1. D. Overbye, *Darkness Visible, Finally: Astronomers Capture First Ever Image of a Black Hole* (New York Times, New York, 2019)
2. M. Lynch, S. Woolgar, *Representation in Scientific Practice* (MIT Press, Cambridge, 1990), p.365
3. C. Coopmans et al. (eds.), *Representation in Scientific Practice Revisited* (MIT Press, Cambridge, 2014)
4. J.C. Le Coze, Visualising safety, in *Safety Research: Evolutions, Challenges and New Directions*, eds. by J.C. Le Coze, T. Reiman (Taylor and Francis, New York, 2013)

---

[6] In the context of environment and climate, the expression is associated with Agenda 21, a non-binding action plan of the United Nations on sustainable development, resulting from the UN Conference on Environment and Development in Brazil in 1992.

5. J.C. Le Coze, New models for new times: an anti-dualist move. Saf. Sci. **59**, 200–218 (2013)
6. B. Latour, Visualization and Cognition: Thinking With Eyes and Hands. Knowl. Soc. Stud. Sociol. Cult. Past Present **6**, 1–40 (1986)
7. C. Jacob, *The Sovereign Map: Theoretical Approaches in Cartography Throughout History* (University of Chicago Press, Chicago, 2006)
8. D.I. Mendeleev, *Mendeleev on the Periodic Law: Selected Writings, 1869–1905* (Courier Corporation, New York, 2013)
9. C. Perrow, *Normal Accidents: Living with High-Risk Technologies* (Princeton University Press, Princeton, 1984)
10. J. Reason, *Managing the Risks of Organizational Accidents* (Ashgate, Aldershot, 1997)
11. J. Reason, *Human Error* (Cambridge University Press, Cambridge, 1990)
12. D.D. Woods et al., *Behind Human Error: Cognitive Systems, Computers and Hindsight* (Dayton Univ Research Inst (Urdi) OH, Dayton, 1994)
13. J. Reason, The identification of latent organizational failures in complex systems, in *Verification and Validation of Complex Systems: Human Factors Issues* (Springer, New York, 1993), pp. 223–237
14. E. Hollnagel, *Barriers and Accident Prevention* (Ashgate, Aldershot, 2004), p.226
15. E. Hollnagel, *Safety-I and Safety-II: The Past and Future of Safety Management* (CRC Press, Boca Raton, 2018)
16. D.D. Woods et al., *Behind Human Error*, 2nd edn. (Ashgate Publishing Ltd., Aldershot, 2010)
17. J. Rasmussen, Risk management in a dynamic society: a modelling problem. Saf. Sci. **27**(2–3), 183–213 (1997)
18. E. Hollnagel et al. Analysis of Comair flight 5191 with the functional resonance accident model, in *The Eighth International Symposium of the Australian Aviation Psychology Association* (Sydney, 2008)
19. E. Hollnagel, *FRAM: The Functional Resonance Analysis Method: Modelling Complex Socio-Technical Systems* (Ashgate Publishing Ltd., New York, 2012)
20. E. Hollnagel, *Understanding accidents-from root causes to performance variability*, in *Proceedings of the IEEE 7th Conference on Human Factors and Power Plants* (IEEE, New York, 2002)
21. R.I. Cook, D.D. Woods, *Operating at the Sharp End: The Complexity of Human Error* (1994)
22. R. Flin, P. O'Connor, M. Crichton, *Safety at the Sharp End. A Guide to Nontechnical Skills* (Ashgate, Farnham, 2008)
23. C. Nemeth et al., Getting to the point: developing IT for the sharp end of healthcare. J. Biomed. Inform. **38**(1), 18–25 (2005)
24. J. Reason, E. Hollnagel, J. Paries, Revisiting the Swiss cheese model of accidents. J. Clin. Eng. **27**(4), 110–115 (2006)
25. J. Braithwaite, R.L. Wears, E. Hollnagel, *Resilient Health Care, Volume 3: Reconciling Work-as-Imagined and Work-as-Done* (CRC Press, Boca Raton, 2016)
26. E. Hollnagel, Why is work-as-imagined different from work-as-done?, in *Resilient Health Care, Volume 2: The Resilience of Everyday Clinical Work*, eds. by R.L. Wears, E. Hollnagel, J. Braithwaite (Farnham, Ashgate, 2015)
27. D. Nathanael, N. Marmaras, Work practices and prescription: a key issue for organizational resilience, in *Resilience Engineering Perspectives: Remaining Sensitive to the Possibility of Failure*, eds. by E. Hollnagel, C.P. Nemeth, S. Dekker, (Aldershot, Ashgate, 2008)
28. B. Latour, *Science in Action: How to Follow Scientists and Engineers Through Society* (Open University Press, Milton Keynes, 1987), p.274

Open Access This chapter is licensed under the terms of the Creative Commons Attribution 4.0 International License (http://creativecommons.org/licenses/by/4.0/), which permits use, sharing, adaptation, distribution and reproduction in any medium or format, as long as you give appropriate credit to the original author(s) and the source, provide a link to the Creative Commons license and indicate if changes were made.

The images or other third party material in this chapter are included in the chapter's Creative Commons license, unless indicated otherwise in a credit line to the material. If material is not included in the chapter's Creative Commons license and your intended use is not permitted by statutory regulation or exceeds the permitted use, you will need to obtain permission directly from the copyright holder.

Chapter 6
Visualising for Safety or Visualisation of Safety?

Erik Hollnagel

Safety is defined and measured more by its absence than by its presence.
Reason [1]

Abstract This chapter considers whether it is possible to visualise safety as a word, a construct, or a concept. It analyses both the instrumental approach of visualising *for* safety (the use of visual means as a help to make systems safe) and the ontological issue of visualisation *of* safety (the use of visual means to show what safety is). It is suggested that the answer depends on whether safety is defined as the absence of unacceptable outcomes (Safety-I) or as the presence of acceptable outcomes (Safety-II).

Keywords Visualisation · Safety · Safety-II · Interpretation

The invitation and motivation for this workshop contained the following statement:

> … our understanding of safety as a construct daily enacted by a multitude of artefacts, actors and institutions has never really been conceptualised from the angle of these drawings, pictures, visualisations, images, but also videos or movies.

Showing or explaining something visually, by pictures or graphics, rather than by words, is generally assumed to be a more effective means of communication. It is indeed a common saying that a picture is worth a thousand words.[1] Safety is, however, only one word, so is it also the case that a picture of safety is a thousand times more valuable than the word or concept itself? Leaving this bogus question aside, the underlying issue is what, if anything is gained by substituting a word, a construct, or a concept by a picture.

---

[1] https://www.phrases.org.uk/meanings/a-picture-is-worth-a-thousand-words.html.

E. Hollnagel (✉)
Professor (Emeritus), Linköping University, Linköping, Sweden
e-mail: sensei@safetysynthesis.com

© The Author(s) 2023
J.-C. Le Coze and T. Reiman (eds.), *Visualising Safety, an Exploration*,
SpringerBriefs in Safety Management, https://doi.org/10.1007/978-3-031-33786-4_6

In this commentary I propose to consider whether it in fact is possible to visualise safety as a word, a construct, or a concept. I will do that by considering two possible interpretations of what the invitation refers to, namely visualising *for* safety and visualising *of* safety. Visualisation *for* safety refers to the use of visual techniques or visual communication to assist in making systems and system performance safe. (This immediately begs the question of what "safe" actually means, as pointed out by the epigraph and as discussed later in this commentary.) In other words, visualising or visualisation with the purpose of promoting whatever processes or behaviours that are believed to support whatever safety is, to influence the behaviour of people (so that they behave "safely") and to present an understanding of how something has happened or could happen (as in accident and risk analyses). It is the issue of visualising what should be done—although in practise it more often is what should *not* be done—to ensure that safety, or a state of safety, is maintained. This obviously also includes visualising what happens when safety is absent, such as copious pictures that show the consequences of accidents, etc.

Conversely, visualisation *of* safety refers to the use of visual techniques to show what safety is or means, assuming that there is something called safety and that that something can be shown. With the risk of being pretentious I might also suggest that visualising *for* safety addresses instrumental problems while visualising *of* safety addresses metaphysical, or perhaps even ontological, problems. To relieve any suspense the reader may have I can already now reveal that the second interpretation in my view is impossible.

6.1 Visualising for Safety

In the case of visualising for safety, the examples are many and varied as shown by the invitation. Although this may appear perplexing, it is nevertheless possible to assign the examples of visualising to a relatively small number of categories. The proposal here is to distinguish between visualising the *outcomes* of safety, visualising the *mechanisms* of safety or how such outcomes can come about, and finally visualising the *safety shaping factors*—what can or should be done to ensure that the specified outcomes occur. Other categorisations may, of course, also be possible.

6.1.1 Visualisation of Safety Outcomes

The visualisation of outcomes are pictures, naturalistic or symbolic, of the consequences of incidents and accidents ranging from the benign or even humorous to the gruesome. From a Safety-I perspective, their purpose is presumably to make the viewer aware of what could happen if care is not taken or if rules and regulations are breached or neglected. Ideally, this should then trigger some kind of avoidance behaviour, in the sense that people will try to avoid these outcomes. Visualisation of outcomes can also be used as an inducement to change behaviour generally, for instance as in the gory pictures shown on cigarette packages in many countries.

In addition to showing possible outcomes, visualisation has also typically been used to show the distribution of various outcomes. The most famous rendering of that is undoubtedly Heinrich's accident pyramid and the iceberg model. The accident pyramid illustrates the advantages as well as the disadvantages of visualising a possible relationship between different types of outcomes. On the one hand, it is easy to understand and use as a reference, specifically when precise ratios of outcome types are assigned, but on the other hand it also suggests a causal relationship that neither exists—or at least has never been proven—and furthermore never was intended (cf., [2]).

6.1.2 Visualisation of Safety Mechanisms

The visualisation of safety mechanisms or of how outcomes happen is best represented by the graphical renderings of accident models, ranging from Heinrich's Domino model and Ishikawa diagrams to Leveson's STAMP, with the Swiss cheese model and the Bow-tie in between. In these cases, the graphical models of how accidents happen are really worth a thousand words, or even more, since a corresponding text would be quite lengthy. (To be fair, the description of the Domino model provided by [3] used only 388 words. But later and more complicated models most likely exceed the 1000 word limit).

The main problem with accident models as a visualisation of safety mechanisms is that they do not visualise safety at all—quite apart from the uncertain epistemic status of what a causal mechanism is [4]. As the epigraph states, safety is defined by its absence rather than its presence—or as a dynamic non-event [5]. An accident is a consequence of the absence of safety, in part or in whole, and a visualisation of an accident either of the outcome or of the way it happens can therefore not be considered a visualisation of safety.

6.1.3 Visualisation of Safety Shaping Factors

If safety is defined by the absence of accidents, then the ways in which accidents can be prevented or avoided must clearly be accepted as visualisations of safety. Examples of this can easily be found ranging from simple matters such as holding on to handrails when using a staircase [6] to a flowchart for maintenance of power transformers (accompanied by about 100 pages of text!).

More generally accidents can be avoided by putting in some barriers which therefore are assumed to serve as safety shaping factors. The purpose of a barrier is to hinder access or passage either in a direct physical sense or in a more metaphorical sense. Barriers, or barrier systems, can be characterised as either physical or material, functional (active or dynamic), symbolic, or incorporeal. (An incorporeal barrier lacks material form or substance in the situations where it is applied and instead depends on the knowledge of the user in order to achieve its purpose, cf., [7].) The three first types of barrier systems rely on visualisation in the sense that they are required to be seen. (It may be argued that a physical barrier system such as a wall will achieve its purpose even if it cannot be seen, but in practice that rarely happens. In fact, when someone bumps into a wall it will most likely be classified as an accident itself).

6.2 Visualising of Safety

The visualisation of safety obviously requires a definition of the essence of safety or of what safety actually is. There is little help to find in the common definitions of safety, which generally equate safety with the relative—or even absolute—freedom from danger, risk, or threat of harm, injury, or loss to personnel and/or property [1, 8]. This is also the essence of the epigraph. (All such definitions reflect the etymology of the English word "safe", which comes from the French word *sauf* that means both "without" and "unharmed.") By using definitions such as these as a starting point, the problem in effect becomes how to visualise nothing. Even if it somehow was possible to visualise nothing, there would be nothing to see, hence no help to understand what safety is.

The concern for the meaning of safety was part of the discussions that led to the formulation of resilience engineering [9]. This later developed into the proposal that it was possible to consider two different ways of interpreting safety which were called Safety-I and Safety-II. According to a Safety-I perspective, safety is defined as a condition where as little as possible goes wrong hence as being *without* unacceptable outcomes. According to a Safety-II perspective, safety is defined as a condition where as much as possible goes well hence as being *with* acceptable outcomes. There is therefore something to visualise but since there is little tradition for doing that, the problem is what should be shown. Perhaps it is what affords safety, rather than what safety is?

6.3 Conclusions

The challenges are obviously not the same for the visualisation of Safety-I and the visualisation of Safety-II, cf. Table 6.1. For the visualisation of Safety-I some of the answers have been given above (at least according to my interpretation). Visualising *for* Safety-I is clearly possible but visualising *of* Safety-I is not. Visualising *for* Safety-II is also possible, although we then need to reconsider how we best show what the outcomes are, how they come about, and how they can be furthered or facilitated— rather than prevented. Visualising *of* Safety-II remains contentious. Even though

Table 6.1 Differences in visualisation for Safety-I and Safety-II

	Safety as the absence of negative outcomes (Safety-I)	Safety as the presence of positive outcomes (Safety-II)
Visualisation of outcomes	Illustrations (realistic/iconic/symbolic) of negative outcomes. Representations of relationships between outcome categories	Illustrations of positive outcomes
Visualisation of mechanisms	Accident models (typically causal and linear)	Models of emergent outcomes and functional couplings
Visualisation of outcome shaping factors	Guidance and instructions; visualisation of barriers	Advice, guidance and instructions

Safety-II can be associated with the presence of something, it is still an elusive concept. In practice, the differences between the two perspectives do not matter much, since the overriding concern is how we best manage or control the processes that lead to specific outcomes rather than how we manage or control safety as such.

6.3.1 Visualisation as a Means to an End

Since there are two significantly different interpretations of what safety is, any analysis of or suggestion for how to visualise safety should recognise the plurality of interpretations and clearly refer to either one or the other rather than to "safety" in general. Visualising safety, whether as Safety-I or Safety-II, should furthermore not be a purpose in itself, but a means to achieve a purpose. Visualisations of safety fall into the category of artefacts in Schein's [10] model of organisational culture, hence must be seen in conjunction with the espoused values and shared basic assumption that also determine performance. While it is beyond doubt that various forms of visualisation can be useful to ensure that work and working environments function as intended—whether in relation to safety, quality, or something else—the pros and cons of visualisation should always be considered relative to the specific purpose. Visualising something is not a magical way to make a diffuse idea intelligible.

References

1. J. Reason, Safety paradoxes and safety culture. Inj. Control Saf. Promot. **7**(1), 3–14 (2000)
2. E. Hollnagel, *Safety-I and Safety-II: The Past and Future of Safety Management* (Ashgate, Farnham, 2014)
3. H.W. Heinrich, *Industrial Accident Prevention: A Scientific Approach*, 4th edn. (McGraw-Hill, New York, 1959)
4. L.J. Matthews, J. Tabery, Mechanisms and the metaphysics of causation, in *The Routledge Handbook of Mechanisms and Mechanical Philosophy*, eds. by S. Glennan, P. Illary (Routledge, London, 2018)
5. K.E. Weick, Organizing for transient reliability: the production of dynamic non-events. J. Contingen. Crisis Manag. **19**(1), 21–27 (2011)
6. J. Cohen, H.H. Cohen, Hold on! An observational study of staircase handrail use. Proc. HFES Ann. Meet. **45**(20), 1502–1506 (2001). https://doi.org/10.1177/154193120104502013
7. E. Hollnagel, *Barriers and Accident Prevention* (Ashgate Publishing Limited, Aldershot, 2004)
8. G.I. Zwetsloot, M. Aaltonen, J.L. Wybo, J. Saari, P. Kines, R.O. De Beeck, The case for research into the zero accident vision. Saf. Sci. **58**, 41–48 (2013)
9. E. Hollnagel, D.D. Woods, N.D. Leveson, *Resilience Engineering: Concepts and Precepts* (Ashgate, Aldershot, 2006)
10. E. Schein, *Organizational Culture and Leadership: A Dynamic View* (Jossey-Bass, San Francisco, 1992)

Open Access This chapter is licensed under the terms of the Creative Commons Attribution 4.0 International License (http://creativecommons.org/licenses/by/4.0/), which permits use, sharing, adaptation, distribution and reproduction in any medium or format, as long as you give appropriate credit to the original author(s) and the source, provide a link to the Creative Commons license and indicate if changes were made.

The images or other third party material in this chapter are included in the chapter's Creative Commons license, unless indicated otherwise in a credit line to the material. If material is not included in the chapter's Creative Commons license and your intended use is not permitted by statutory regulation or exceeds the permitted use, you will need to obtain permission directly from the copyright holder.

Chapter 7
Visualizing Complex Industrial Operations Through the Lens of Functional Signatures

Doug Smith, Brian Veitch, and Arash Fassihozzaman Langroudi

Abstract In this chapter, the concept of functional signatures is presented as a way to understand complex industrial operations. Visualization of functional signatures can be used to improve tractability of complex operations, which can be valuable for safety analysis. Two techniques for visualizing functional signatures are presented: (1) cyclic functional signatures and (2) linear functional signatures. Both techniques are seen as valuable and selection of technique can be left to user preference. The two visualization techniques are demonstrated through an application of an ice management operation for an offshore petroleum installation performed in a simulated ship environment.

Keywords Visualisation · Functional signatures · System modelling · FRAM

7.1 Introduction

Visualization can enrich the understanding of industrial safety, especially for complex industrial operations. In complex industrial operations, the work is often under-specified. The success of such operations relies on local adjustments by workers to account for the under-specification. Furthermore, these local adjustments can cause chaotic system behaviours over time and make it difficult for safety and risk assessors to imagine how outcomes might emerge from the operations. The difficulty in foreseeing outcomes and anticipating system behaviours can make the operation seem opaque or intractable. The opaqueness of the operations can lead to improper diagnosis of safety concerns and, in turn, poor safety management decisions.

D. Smith (✉) · B. Veitch
Faculty of Engineering and Applied Science, Memorial University of Newfoundland, St John's, NL, Canada
e-mail: d.smith@mun.ca

A. F. Langroudi
Department of Naval Architectural Engineering, Memorial University of Newfoundland, St John's, NL, Canada

© The Author(s) 2023
J.-C. Le Coze and T. Reiman (eds.), *Visualising Safety, an Exploration*,
SpringerBriefs in Safety Management, https://doi.org/10.1007/978-3-031-33786-4_7

This operational opaqueness has been evidenced in conventional approaches to safety analysis. One popular approach to safety analysis is to hypothesize about important accident causing factors and examine an operation to obtain an understanding of each factor's significance to accident causation. This approach has been used with some effectiveness, but while it may help identify some of the significant contributors to accidents, it does not provide an understanding of how system behaviours might be affected by making changes to the identified contributor. The factors' interconnection with other operational elements is important information for explaining operational outcomes and implementing safety management decisions. Consider that the accident causing factors that may be of interest to your analysis are part of a larger system that have complex inter-relations that contribute to the system's functionality. It will be difficult to ascertain whether the factors being assessed are also influenced by other factors within the system without somehow examining the system as a whole. This gap can be addressed by adopting a system analysis approach which focuses on understanding the entire system rather than just a few elements of it. Figure 7.1 illustrates how the traditional approach can leave the system seeming opaque and how a system modelling approach can help illuminate the system structure.

The complexities in operational systems can be difficult to imagine and thus having a visualization tool to animate the processes is advantageous. In this chapter, we propose using functional signatures, which is an extension of the functional resonance analysis method (FRAM), to visualize complex operations with the purpose of enhancing learning and informing safety management.

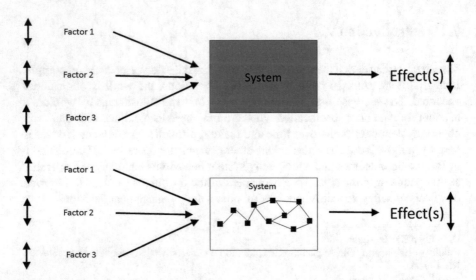

Fig. 7.1 Analysis without system model (top) versus analysis with system model (bottom)

7.2 Background

7.2.1 Fram

The functional resonance analysis method (FRAM) is a system functionality modelling method [1]. The method has two main modelling components: (1) the functions or activities and (2) the variability. The first part of the FRAM involves describing the potential functions that are or could be used to achieve some goal. The functions are work that is done within the system. The basic rule to distinguish functions from other system parameters is to determine whether something is being produced. If there is nothing being produced, there is no work being done, and thus it should not be modelled as a function. Each function has 6 aspects that can be described: (1) input(s), (2) output(s), (3) resource(s), (4) time, (5) control(s) and (6) precondition(s). The aspects help understand the expectations of each function, by explicitly considering what is produced [output(s)], what initiates the function [input(s)], what is required prior to the function's execution [precondition(s)], what is consumed by the function [resource(s)], what constrains the functional process [control(s)] and what ways are the available time to execute the function affected (time). In addition to providing an understanding of how a function might be executed individually, the aspects also provide a mechanism to connect multiple functions and build a functional system. As outputs are produced by functions, they may be utilized by other functions, thus coupling one function's outputs to one or more of the other five aspects for another function. Figure 7.2 illustrates the concept of a functional node and a functional system.

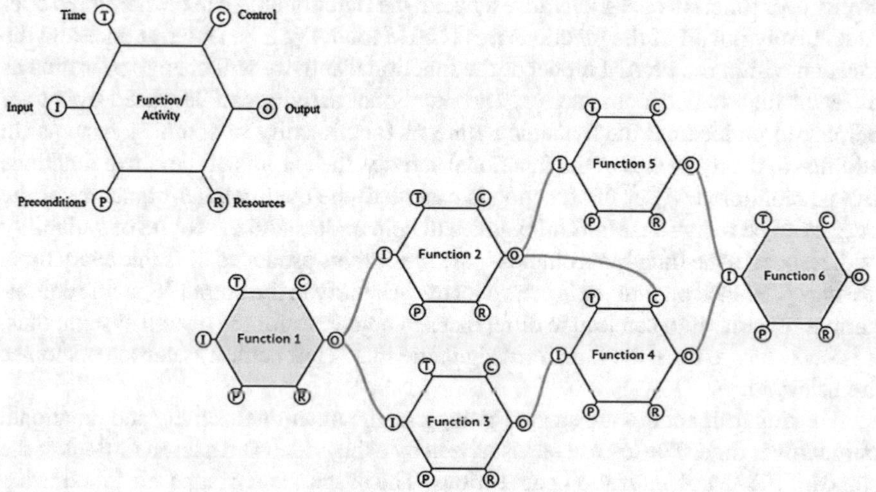

Fig. 7.2 FRAM node (top left) and FRAM model (bottom right)

The second part of the FRAM is to model the variability. The FRAM focuses on modelling the variability in two ways: (1) the variability of individual functional outputs and (2) the coupled variability of multiple functions. The individual functional variability refers to the output of a single function. Individual outputs can vary with respect to time and with respect to precision. High variability functions are not necessarily seen as harmful, in some cases there must be an allowance for high variability to accommodate many possibilities. Similarly, low variability is not necessarily seen as desirable. Once the individual functional variability is characterized, the next step is to understand how certain configurations of variability throughout the system might combine to produce outcomes. The understanding of coupled variability can be used to monitor and manage the functionality of systems.

7.2.2 Functional Signatures

The FRAM is itself a visualization technique. FRAM models can be visualized using the FRAM model visualizer (FMV) which is freely available online [2]. The FMV allows users to visualize their FRAM models, displaying the functions and relationships between them. The model is a representation of the potential functional processes that could be executed to achieve the overall goal of the system. This static visualization of the FRAM model illustrates the first part of the FRAM—describing the system—but does not offer the ability to visualize the variability.

Functional signatures can be used to visualize the variability of operations by way of an extension to the standard FRAM visualization. Functional signatures are recorded accounts of the functional activity and individual functional output variability over time. As operations are executed, the functionality can be monitored over time. Likely not all of the functions in a FRAM model will be active at once and the location within the FRAM model of the functional activity will change over time as different functions become active. The functional activity can be traced over time helping to understand the dynamic nature of functionality in complex systems. In addition to the dynamics of the functional activity, the outputs of the active functions can be monitored. After the function is executed, the quality and/or quantity of the output can be recorded. Monitoring this will help understand the nature of variability with respect to the functional outputs—are the outputs produced the same every time, are they produced on time, or is there a lot of variability in the output? Specific combinations of variability can lead to differences in outcomes for the overall system; thus, it is important to trace the functional signature that is left behind as certain outcomes are achieved.

Functional signatures are created by logging the functional activity and functional outputs over time. The log can be visualized by using video and a few additions to the standard FRAM visualization conventions. The system can first be modelled using the standard FRAM procedures [1]. At this stage, the model represents the potential functional possibilities of the operation. The functional signature—the specific functional activities and functional outputs over time—can then be visualized by (1)

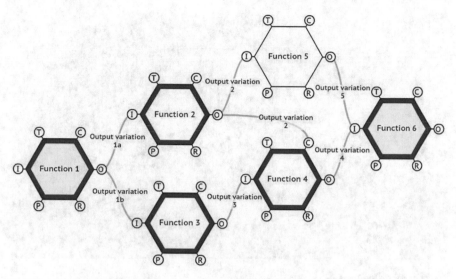

Fig. 7.3 Functional signature with active functions bolded in red

distinguishing active functions from inactive functions usually by using colour and (2) writing the quality and/or quantity of a specific output on the line that connects the active function's output to a downstream function. See Fig. 7.3 for an example of this concept.

7.3 Discussion

The concept of functional signatures is rather simple but the manifestations that occur when monitoring complex operations can be more complicated. The level of complexity will depend on the nature of the operation. Operations that have a large number of functions that are highly connected will generally produce more complex functional signatures than smaller, less connected ones. Other operations that can produce more complex manifestations are operations that are modelled recursively. A recursively modelled operation is an operation that requires constant management in real-time and allow for worker interventions. When modelling an operation recursively, the focus is on modelling the management-intervention process. The management-intervention process is repeated as many times as needed. This means the model could repeat many times while the operation is working towards achieving its goal, thus increasing the complexity.

In order to demonstrate the functional signature concept, an example is used based on ship navigation in a simulated ship environment. A ship simulator for ice management was used to monitor an operation and generate functional signatures. The ship simulator consists of a platform with ship-like controls, situated in the

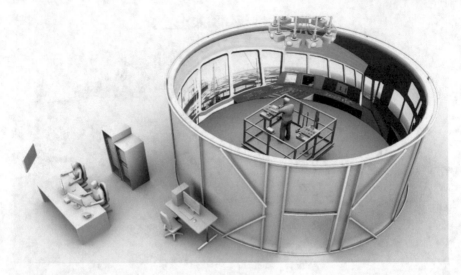

Fig. 7.4 Ship simulator setup

centre of a wraparound projection screen, where users have a 360° view of the environment they are operating in. Figure 7.4 shows the setup for the ship simulator. The ice management operation involved a support vessel (AHTS—anchor-handling tug supply vessel) for a larger "fixed" offshore installation that was surrounded in pack ice. The driver of the support vessel was asked to clear a zone in the pack ice where a lifeboat could be launched. This scenario is outlined in Fig. 7.5. As the driver navigated the support vessel through the pack ice, the functionality of their operation was tracked.

The first step is to create a FRAM model for ship navigation. A model was created via interviews of ship captains. The details of the modelling process can be seen in [3]. However, the model presented here has a reduced scope to minimize the clutter of functions that could not possibly become active in the simulated environment. The FRAM model is displayed in Fig. 7.6. The FRAM model describes how the driver will make assessments of their situation and decide whether or not to change course (intervene) or maintain it. The assessment will be based on their understanding of the ice conditions, where they are positioned in their environment, their vessel parameters and an awareness of a regulatory requirement to keep the speed of this vessel below 3 knots in the ice field. This process is repeated as many times as needed as the driver tries to achieve their goal—clearing ice from the lifeboat area.

By using this FRAM model and monitoring the functionality of the driver during the operation, functional signatures can be created. There are two visualization techniques that can be used to view this type of recursively modelled operation: (1) cyclic functional signatures or (2) linear functional signatures. There is value in both of these visualization techniques and one is not necessarily better than the other. Users may use both visualization techniques as dictated by their preference.

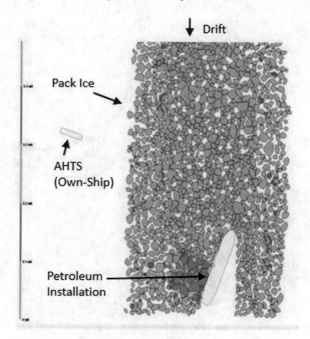

Fig. 7.5 Scenario configuration for the ship simulator

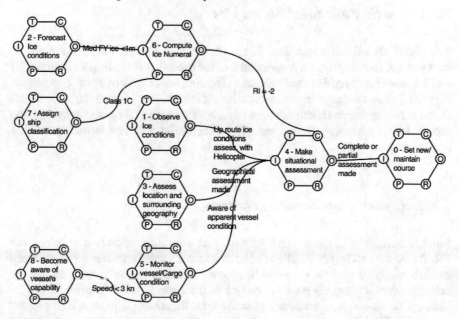

Fig. 7.6 FRAM model for ship navigation

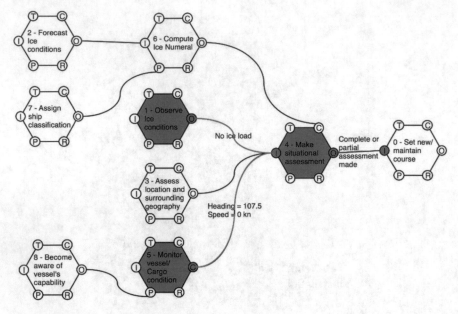

Fig. 7.7 Cyclic functional signature 11 s after beginning

7.3.1 Cyclic Functional Signatures

The cyclic visualization technique keeps the original FRAM model static, in the background, and overlays the functionality of the operation as it animates. Figures 7.7 and 7.8 demonstrate of a cyclic functional signature for a driver of the ship simulator. Figure 7.7 shows the assessment function becoming active after receiving two outputs from the two upstream functions at time equals 11 s. Figure 7.8 shows the cycle being repeated after a heading change was made. The cyclic will then repeat as operator performs the next task.

7.3.2 Linear Functional Signatures

Linear functional signatures march forward with time as the operation is shown. The FRAM model will be copied ahead of itself and the active functions are highlighted as each recursive cycle is shown. Also, there can be an option to only display the active functions when the model is copied in front and hide the inactive functions. This option can hide unnecessary clutter as the functional signature is displayed. Figure 7.9 displays a snapshot of a linear functional signature a driver of the ship simulator with the inactive functions hidden. Figure 7.9 shows another cycle of the model beginning with three active functions after a course change. It is also important to note that the model may appear stretched or compressed with respect to the original

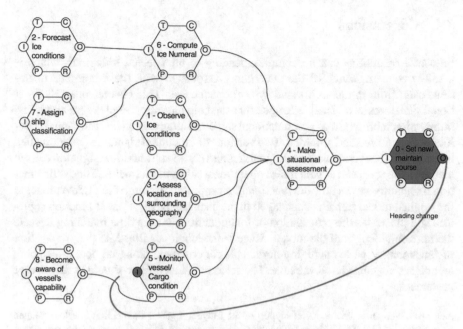

Fig. 7.8 Cyclic functional signature 70 s after beginning

FRAM model layout. The stretching and compression is determined by the time it takes for functions to be completed. Functions that take longer to be completed will appear stretched opposed to functions that are completed quicker.

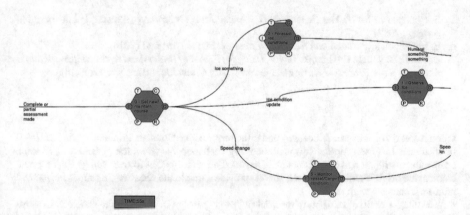

Fig. 7.9 Linear functional signature 55 s after beginning

7.4 Conclusions

Industrial operations can be complex, leading to unexpected outcomes, which is a safety concern. Much of the uncertainty associated with these complex operations stem from intractability and system opaqueness. This chapter presents functional signatures as a visual technique that can help monitor complex operations and improve tractability. Creating a functional signature requires that key functional activities are monitored and recorded. Two versions of functional signatures are presented in this work: (1) cyclic functional signatures and (2) linear functional signatures. Both visualizations are seen as valuable to operational safety assessments. The cyclic functional signature shows how the functionality propagates through the FRAM model as the original model remains unchanged in the background. The linear function signature still illustrates the propagation of functionality but uses time marching to scale the horizontal length of the model. Since both techniques illustrate the propagation of functionality, which is highly intractable for complex industrial operations, both techniques are considered valuable. The selection of technique should be left to user preference.

Acknowledgements The financial support of the Lloyd's Register Foundation is acknowledged with gratitude. Lloyd's Register Foundation helps to protect life and property by supporting engineering-related education, public engagement and the application of research.

References

1. E. Hollnagel, *FRAM: The Functional Resonance Analysis Method* (Ashgate Publishing Ltd., Ashgate, 2012)
2. E. Hollnagel, *The Functional Resonance Analysis Method Manual* (2018)
3. D. Smith, B. Veitch, F. Khan, R. Taylor, *Using the FRAM to Understand Arctic Ship Navigation: Assessing Work Processes During the Exxon Valdez Grounding* (TransNav 12, 2018)

Open Access This chapter is licensed under the terms of the Creative Commons Attribution 4.0 International License (http://creativecommons.org/licenses/by/4.0/), which permits use, sharing, adaptation, distribution and reproduction in any medium or format, as long as you give appropriate credit to the original author(s) and the source, provide a link to the Creative Commons license and indicate if changes were made.

The images or other third party material in this chapter are included in the chapter's Creative Commons license, unless indicated otherwise in a credit line to the material. If material is not included in the chapter's Creative Commons license and your intended use is not permitted by statutory regulation or exceeds the permitted use, you will need to obtain permission directly from the copyright holder.

Chapter 8
Anticipating Risk (and Opportunity): A Control Theoretic Perspective on Visualization and Safety

John M. Flach

Abstract A central challenge in designing stable control systems is to identify the states that must be fed back to enable successful control. The quality of control (including safety) depends on our ability to visualize the state space underlying the functional dynamics of the work being managed. Building concrete visualizations is both a useful tool for knowledge elicitation with domain experts to discover the meaningful functional work constraints that determine this state space, and an essential part of interface design to support safe work in complex systems.

Keywords Visualization · Ecological interface design · Elicitation · Semantic mapping · Systematicity

8.1 Introduction

One of the central challenges in designing stable control systems is to identify the "states" that must be fed back to enable successful control. For successful control, these states must allow a controller to anticipate the future consequences of actions (or inaction). For example, it is impossible to control an inertial vehicle (e.g., a car) with feedback of position only. It is also essential that the velocity also be fed back, since that provides a basis for anticipating future positions. Further, the weighting of position and velocity in order to know when to initiate braking must reflect the dynamic capabilities of your brakes [6]. In other words, distance and velocity feedback and the associated relations to braking dynamics are essential for letting drivers judge safe speeds and following distances and when to initiate braking.

In generalizing this insight to complex, high-dimensional control problems, the conclusion that we draw is that the feedback provided to the controllers of these systems must make the states and patterns of constraint among the high-dimensional states of the systems being controlled salient to decision makers so that they are

J. M. Flach (✉)
Mile Two LLC, Dayton, OH, USA
e-mail: jflach@miletwo.us

© The Author(s) 2023
J.-C. Le Coze and T. Reiman (eds.), *Visualising Safety, an Exploration*,
SpringerBriefs in Safety Management, https://doi.org/10.1007/978-3-031-33786-4_8

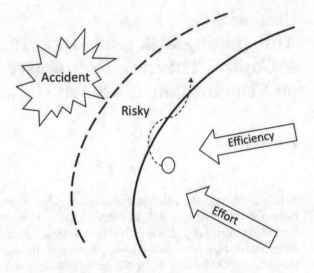

Fig. 8.1 Rasmussen's dynamic safety model showing drift across safety buffers

better able to anticipate the consequences of their decisions and actions. However, in complex open systems the number of potentially relevant state variables can be large and some of these variables may be difficult to measure and specify. One strategy for coping with this challenge is "defence in depth." That is, to specify a tangible (well-specified) boundary or buffer to protect the system from risky situations (e.g., a speed limit). However, as [10] observed, defence in depth solutions are vulnerable to decay over time as people endeavour to increased efficiency and minimize effort (Fig. 8.1). Each time they drift pass the "artificial" constraint its power for influencing behaviour diminishes, increasing the potential for an accident.

The point of Rasmussen's dynamic safety model is that defence in depth is not sufficient. In addition to creating buffers to protect the system from risky situations, it is important to directly face the difficult challenge of making the actual risk boundaries visible (i.e., providing the feedback that operators need to anticipate and avoid dangerous situations). This insight is the motivation for the Ecological Interface Design approach for creating representations or visualizations for safety critical systems (e.g., [2, 11]).

In terms of interface design to support complex work, representations tend to be framed as either geometric analogues or metaphors that have multiple levels of structure to reflect the multiple levels of means-ends constraints associated with complex work (e.g., [2, 5]). For both analogical and metaphorical representations, there are two related principles that are fundamental to the quality of the representation: semantic mapping [3] and systematicity [7].

8.1.1 Semantic Mapping and Systematicity

The semantic mapping principle states that there should be a "one-to-one mapping between the invisible abstract properties of the process and the cues or signs provided by the interface" [11]. This principle emphasizes the importance of correspondence between meaningful properties of the dynamics of the work and properties of the associated analogy or metaphor. The most important properties for anticipating the consequences of action should be the most salient features in the representations. The goal is to help workers to directly "see" the future consequences of their decisions and actions (i.e., affordances—opportunities and risks).

The systematicity principle states that "a system of relations connected by higher-order constraining relations such as causal relations is preferred over one with an equal number of independent matches" [7]. This principle emphasizes the relations across multiple levels of constraint. The proposition is that the functional patterns across levels of constraint should correspond with similarly nested patterns within the representation. Complex work has been modelled as a nested hierarchy of constraints (e.g., [9]). In this context, systematicity reflects the degree to which the structure of the analogue or metaphor corresponds with this nested structure. For example, high orders of constraints (e.g., goals, values, safety) should be reflected in global properties (e.g., global symmetries) and lower orders of constraints (e.g., component interactions) should be mapped to local features nested within more global patterns (e.g., local symmetries).

Woods [13] addresses one aspect of systematicity with the construct of visual momentum. Realizing that for very complex work the work may need to be distributed across multiple display pages, Woods recognized that the work needed to be parsed in a way that preserved both local and global coherence. Using the metaphor of editing film, he discusses multiple techniques for preserving global relations across multiple local display windows. In other words, the parsing of the work across multiple representations must respect the higher-order structural relations underlying the work dynamics. This is consistent with the principle of systematicity.

This is not just important for parsing work for an individual, but the same principles apply to how we distribute information across multiple operators in a distributed work context. In building interfaces for distributed or polycentric control systems the parsing of the work must preserve the multilevel structure of functional constraints—so that, individuals see what is meaningful locally in relation to more global common goals and values. This is essential for achieving coordinated control.

I would like to make the case that these two principles (semantic mapping and systematicity) are fundamental to all forms of representations—computer interfaces, internal mental models created through training, posters, and movies. For example, just as an interface can be evaluated in terms of the mapping of work semantics and the appropriate layering of information to reflect levels of constraint associated with the work dynamics, so too, can a movie be evaluated in terms of the structure of the narrative and how local patterns of events fit within more global themes that reflects how local events relate to higher-order values associated with safety.

Returning to the control theoretic context, an important implication of this is that a comprehensive work analysis is essential to designing appropriate representations/ visualizations. That is, the goal of work analysis is to provide a model of the "state" space that reflects the underlying state variables and the patterns of constraint among them. While I am tempted to say that this is a prerequisite for designing effective visualizations, experience tells me rather that it is a co-requisite. This reflects my experience that visualizations themselves are critical to the process of work analysis. In 30 years of working on designing interfaces for sociotechnical systems, I have found that building concrete visualizations (e.g., wireframe interfaces) can be essential for knowledge elicitation with domain experts to discover the meaningful functional work constraints. The first interface concept generated on the basis of extensive work analysis is rarely sufficient. However, the initial concepts can be extremely valuable in engaging domain experts in a participatory design process. Often, the interactions in assessing and evaluating initial interface designs help us and the domain experts to gain a heightened awareness of the information feedback that is necessary for effective control. Thus, concrete representations can be essential for creating the common ground essential to multidisciplinary collaborative design.

8.1.2 Some Examples

Figure 8.2 shows four examples of ecological interfaces. Although the representations are very different at the surface level, each was designed to explicate the links between system states, actions and risk. For example, Vicente's [12] DURESS interface explicitly links the fluid flows through a feedwater control system with the mass and energy targets and the ultimate constraints on safety associated with the balancing mass and energy. Amelink et al. [1] Total Energy Reference Path interface is designed to help pilots see and understand the relation between manipulations of their controls (e.g., stick and throttle) and a safe balance between kinetic and potential energy while landing. The Cardiac Consultant interface [8] is designed to explicate the links between various clinical and behaviour measures and the risk of cardiovascular disease. And finally, the RAPTOR interface [4] includes a Force Ratio display as an explicit indication of the risks associated with military engagements.

8.2 Summary

In sum, the point is not to eliminate defence in depth protections, but to understand that defence in depth protections alone will often not be sufficient. It is not a question of *either* defence in depth *or* representation design. Rather, for safety critical systems it is dangerous to rely on either alone. We need both. For example, in driving speed limits and lane marking provide important safety buffers, but safety can be further improved by adding additional feedback about actual risks (e.g., blind spot displays).

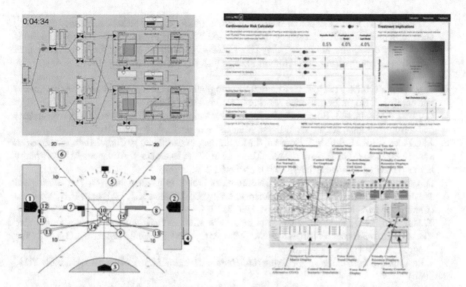

Fig. 8.2 Ecological interfaces are designed to help operators to better see the factors that impact safe operations

Thus, in addition to safety buffers, it is important to face up to the challenge of providing the feedback that will help operators to anticipate and respond to the actual boundaries to safe operations. The essential point is that control requires the ability to anticipate the consequences of decisions and actions. For safety, this means the ability to anticipate risks in time to take action to avoid or mitigate them. The physical and mental visualizations that people use to assess the "state" of the system will determine their ability to anticipate risks. Today, designers have a wide range of opportunities to shape these visualizations through many different media. However, independent of the medium, the quality of the visualization will depend fundamentally on the mapping of the visualization to the functional semantics of the work being performed. The meaningful functional constraints must be salient, and the organization of the constraints must systematically correspond to multilevel relations that shape the functional work dynamics. In essence, the quality of control (including safety) will ultimately depend on our ability to visualize the state space underlying the functional dynamics of the work being managed.

References

1. H.J.M. Amelink, M. Mulder, M.M. van Paassen, J.M. Flach, Theoretical foundations for total energy-based perspective flight-path displays for aircraft guidance. Int. J. Aviat. Psychol. **15**, 205–231 (2005)
2. K.B. Bennett, J.M. Flach, *Display and Interface Design: Subtle Science, Exact Art. Boca Raton* (CRC Press, FL), ISBN-13: 978-1420064384 (2011)

3. K. Bennett, J.M. Flach, Graphical displays: implications for divided attention, focused attention, and problem solving. Hum. Factors, **34**(5), 513–533 (1992) Reprinted in N.J. Cooke, E. Salas (eds.), *Best of Human Factors: Thirty Classic Contributions to Human Factors/Ergonomics Science and Engineering*. The Human Factors and Ergonomics Society, Santa Monica, CA (2008)
4. K.B. Bennett, S.M. Posey, L.G. Shattuck, Ecological interface design for military command and control. J. Cogn. Eng. Decis. Making **2**(4), 349–385 (2008)
5. C. Borst, J.M. Flach, J. Ellerbroek, Beyond ecological interface design: lessons from concerns and misconceptions. IEEE: Trans. Syst. Man Cybern. **45**(2), 164–175 (2015)
6. J.M. Flach, M.R.H. Smith, T. Stanard, S.M. Dittman, in *Collision: getting them under control*, eds. by H. Hecht, G.J.P. Savelsbergh. Theories of Time to Contact. Advances in Psychology Series (Elsevier, North-Holland, 2004), pp 67–91
7. D. Gentner, B. Bowdle, P. Wolff, C. Boronat, Metaphor is like Analogy, in *The Analogical Mind: Perspectives from Cognitive Science*. ed. by D. Centner, K.J. Holyoak, B.N. Kokinov (Cambridge MA, MIT Press, 2001), pp.199–253
8. T. McEwen, J.M. Flach, N. Elder, in *Interfaces to Medical Information Systems: Supporting Evidence-Based Practice*. IEEE: Systems, Man, & Cybernetics Annual Meeting, Oct 5–8 (San Diego, CA, 2014), pp. 341–346
9. J. Rasmussen, *Information Processing and Human-Machine Interaction* (Elsevier, New York, 1986)
10. J. Rasmussen, Risk management in a dynamic society: a modelling problem. Saf. Sci. **27**, 183–213 (1997)
11. J. Rasmussen, K.J. Vicente, Coping with system error through system design: implications for ecological interface design. Int. J. Man Mach. Stud. **31**, 517–534 (1989)
12. K.J. Vicente, *Cognitive Work Analysis* (Erlbaum, Mahwah, NJ, 1999)
13. D.D. Woods, Visual momentum: a concept to improve the cognitive coupling of person and computer. Int. J. Man Mach. Stud. **21**, 229–244 (1991)

Open Access This chapter is licensed under the terms of the Creative Commons Attribution 4.0 International License (http://creativecommons.org/licenses/by/4.0/), which permits use, sharing, adaptation, distribution and reproduction in any medium or format, as long as you give appropriate credit to the original author(s) and the source, provide a link to the Creative Commons license and indicate if changes were made.

The images or other third party material in this chapter are included in the chapter's Creative Commons license, unless indicated otherwise in a credit line to the material. If material is not included in the chapter's Creative Commons license and your intended use is not permitted by statutory regulation or exceeds the permitted use, you will need to obtain permission directly from the copyright holder.

Chapter 9
Occupational Safety in Revamping Operations: Visualising Spaces to Monitor Uncertainty

Charles Stoessel and Raluca Ciobanu

Abstract Deployed from 2009, the purpose of the Design Safety approach is, "to put in place the technical and/or organisational resources to reduce the risks incurred by those involved during the construction phase" (Mbaye and Saliou 2014). We seek to understand the role of spatial visualisation and workplace situations in risk prevention. A field study was conducted over a three-year period, suggesting that Design Safety involves strategic decision-making during upstream project phases, combined with operational decision-making during downstream phases, each with an impact on safety. In this context, the work of designers should be supported by specific artefacts (databases of photographs, technical documentation, 3D visualisation, etc.) to give them a better understanding of the workplace situation and visibility in terms of the risks incurred by the workers.

Keywords Design safety · Risk visualisation · Collective activity · Risk prevention · Occupational health and safety

9.1 Introduction

This study depicts the integration of occupational safety (prevention of injuries, for the workers, that should be distinguished from industrial safety, dealing with the risk of high-scale accidents). Occupational safety is seen throughout the following processes: constructing ("new build"), decommissioning, but also "revamping". These revamp operations are less studied by the occupational safety and project management literature, in particular the implementation of new equipment/replacement of older equipment in the operating units. These operations may

C. Stoessel (✉)
Opus Citatum and Paris-Dauphine University, Paris, France
e-mail: charles.stoessel@opus-citatum.com

R. Ciobanu
EDF R&D, Palaiseau, France
e-mail: raluca.ciobanu@edf.fr

© The Author(s) 2023
J.-C. Le Coze and T. Reiman (eds.), *Visualising Safety, an Exploration*,
SpringerBriefs in Safety Management, https://doi.org/10.1007/978-3-031-33786-4_9

be complicated because engineers don't start from a "blank page" such as in new builds. On the contrary, they have to take into account the existing activity and the previous history of the operating unit to "insert the transplant" properly. This is a key reason why visualisation will play a major role in this configuration.

The present study aims at producing concepts and operational tools in order to ensure the Prevention through Design process. Thanks to the commitment of the Engineering Division (security experts, projects managers, directors…), a large amount of data have been collected (interviews, technical observations, meetings participation…) during a 3-year research project (2013–2016) in an electrical company which performs high-scale maintenance operations such as replacement of steam generators. The study focuses on the visualisation of the work environment (a steam generator pipe) in classical engineering functional diagrams, represented by the 3D model, and the real working environment experienced/seen by the worker.

These gigantic projects have to take into account classical project management objectives such as time, cost, and quality. In addition, high-risk organisations have to monitor industrial safety and specific industry, such as nuclear power plants, have to minimise workers' radioactivity exposure. Nevertheless, all these additive objectives must be managed simultaneously. Innovation can create new technical solutions in order to ensure project effectiveness (time/cost) while deploying construction techniques that prevent the workers from injuries and fatalities. The right equipment can save lives: it has to be purchased early enough. In order to set the right equipment for each work situation, the designers should be able to visualise what will be the actual work condition encountered by the workers.

The research shows how designer's training must integer the use of visualising artefacts such as work situations photos databases, commented and related to context and work stories by experimented trainers. The outputs will also reveal the importance of using visualisation artefacts such as 3D models that clearly help the people designing and planning the operations to get to a better representation of the complexity and varieties of work situations.

9.2 Design Safety and Risk Visualisation

Prevention through Design implies training the designer in experiencing the field difficulties that can be encountered by operations teams, in order to bring them to take into account this occupational safety dimension while designing the early stages of the project. Training the designer should help him understand the difficulties occurring during the construction phase notably through the visualisation of "real work" complexity. For example, working at height is often invisible from the designer's perspective, but may be a real hazard and a complex situation for the worker involved in this very situation. The design safety objective requires building cooperation within the organisation, implementing dialogue between engineering and risk prevention expertise. Risk visualisation tools can play a major role in this respect.

9.2.1 Active Participation and Permanent Reconfiguration of a Working Group

The notion of design covers all work preparation activities, technical processing of files by design engineers, works planning, budgetary management of projects. The objectives of design safety are to (i) improve the safety of those involved from the design phase, via the detection of potentially hazardous workplace situations (ii) ensure optimal work site safety by taking into account different facets of the activity (technical, ethical, contractual management). It is essential to take all these facets into account to ensure the success of design safety integration projects. As such, design safety relies on collective activity (engineers, design offices, prevention experts, project managers, managers), more specifically on a pragmatist representation of workplace situations [4, 19–21, 27, 28].

9.2.2 Interaction of Workspaces and Tools for Integrated Prevention Purposes

Some of the research emphasises the use of the work environment as a resource to organise the current and future actions of professionals [5, 22, 31]. Not only must design engineers find their bearings in this organisational space (geographical area in production centres), they must also project themselves in the relational space by imagining interactions between stakeholders (project coordinator, work supervisor, workers, etc.) who play an important role in their operational intervention [23]. In this context, space can serve as support or constitute a difficulty for their activity [24, 25].

For the purposes of this chapter, we investigate how design engineers perceive specific design safety issues pertaining to the revamping of existing facilities. While their role and the scope of their actions as part of the Design Safety approach are somehow predefined by the company's process and quality policy, they are not limited to these requirements. How do these engineers, often recently hired within engineering centres, perceive the scope of design safety? Which tools do they rely on to analyse the risks associated with operations? How do they respond to the increasingly demanding reliability requirements in terms of risk analyses and, more generally, workers' activity?

9.2.3 From Situated Action to Risk Visualisation

Numerous academic studies highlight the importance of a pragmatist design of work-place situations and situated decision-making [4, 19–21, 27, 28]: "an analysis as detailed as possible of future workplace situations from the design stage helps iden-tify any risks, with a view to eliminating them or, failing this, reducing them but in any event controlling them" [26], p. 7. Accordingly, the major challenge of the safety of those involved in the design phase is to provide designers with the means to envisage safety in very concrete terms from an early stage, even though it will only come into play years later, during the construction phase. In keeping with research in design ergonomics [8, 9, 11, 13, 18, 35], the authors often emphasise the amount of autonomy required to perform these activities.

9.2.4 Safety Integration and Digital Simulations

While a number of standards describe how to provide for safety integration and work-station ergonomics, they must be supplemented by digital simulation tools. These tools help visualise the design of workspaces so as to rapidly integrate good safety practices. Within modelled work spaces, the use of digital "dummies", integrated into work environments represented via CAD software (Computer Aided Design), improves the visualisation of the users' workplace situations in order to improve risk prevention.

Designers can therefore be instrumental in preventing risks to workers (new struc-tures and, by extension, revamping of existing structures), as long as they are trained and have a thorough understanding of the manufacturers' requirements [3]. In addi-tion, designers argue that they rarely benefit from appropriate initial training as well as tools allowing them to take personnel safety into account [16]. It is therefore advis-able to model the reality, whether physiological or psychological processes, potential accidents, man–machine interactions, by using three-dimensional "dummies" and simulating possible interactions between users and the system [12], p. 63.

Our literature review suggests that the design safety project should integrate skills development for prevention experts and designers, notably via training. Educational methods must be suited to the necessary "visualisation" of situations, the develop-ment of safety skills which are often learned through direct and situated professional socialisation, in contact with working instruments.

9.3 Methodology

This research involved an inductive survey methodology, allowing actual cases from the field to guide the discussion and adjust initial working assumptions. Once we finished collecting data, we analysed a large number of internal documents relating to the design safety approach, activity observation reports and interviews, as well as field notes.

The study was conducted within two engineering centres and two production sites. In total, we conducted around twelve semi-structured individual interviews which lasted one hour on average, a collective 4 h interview with Safety-Design engineers, 4 observation sessions of safety and operational meetings, as well as spending several days observing the working activity of project teams on sites. The jobs encountered featured safety project managers (internal and service providers), safety controllers, safety-design engineers, field surveillance staff, on-site revamping project managers, etc.

9.4 Results

The main results highlight the current limitations of the risk analyses produced by designers and encourage us to develop tools to visualise the actual configurations of work sites, with a view to addressing this socio-professional discrepancy.

9.4.1 Risk Analysis: From the Designer to the Worker

The field survey, conducted on revamping sites, provides a nuanced picture of the use of risk analyses carried out in engineering centres. Some interviewees feel that these safety risk analyses establish the major principles but fail to go into enough detail to be directly usable downstream of the process. At the moment, risk analyses received in files are not really used on work sites. This means that formal design safety processes are faced with a major difficulty: the ability to "project" from the design situation to the site implementation situation.

Those deployed on the sites are not surprised by this lack of specificity in design risk analyses. Engineering centres find it difficult to benefit from the local, contextual, situated information required for finer risk analyses. The job of those working on the ground is precisely to adapt a generic design file to the local context. Therefore, an important operational avenue would be to increase the amount of information available in engineering centres.

In concrete terms, designers cannot always avail themselves of technical plans and sufficiently reliable and exhaustive photographic databases to fully project into the local context of the actual work site. A possible improvement would therefore be to increase the knowledge available in engineering centres on the "life of work sites" and the actual working conditions of those involved in factories (e.g., time to access the site, joint activity, rescheduling, etc.).

9.4.2 Good Visualisation Practice on Project "CCR43"

Revamping project "CCR43"[1] was examined during the revamping "integration" phase, on the production site, i.e., during the implementation phase. Project CCR43 is part of a large-scale "VGR" programme which includes a number of VGR work sites at several production locations as well as several CCR43 work sites, also at different production locations. In the plants concerned, work is carried out in a severely restricted environment. There are many risks and space is extremely limited. This is why it is important to visualise the working area to schedule works as accurately as possible. The operation is carried out in a "pillbox", a multi-storey room more than twenty metres high, with each storey just a few square metres in area.

Of particular importance for risk prevention and visualisation, we observed that the working teams frequently referred to a three-dimensional model of the work site. The "Elbow" technical object, which must be replaced on the CCR43 work site, can thus be easily visualised by the team. In the progression of the visualisation artefacts, this technical object is successively represented using three different visualisation artefacts, functional diagram, 3D model and on-site photo.

The team also uses photographic representations of the "reality" (or rather a portion thereof) of the CCR43 work site and presents a particularly complex working context for the safety of those involved (work at a height, numerous cables, very cramped spaces, multiple pieces of equipment, etc.).

This type of visualisation tool (including photographs, diagrams, sketches, plans, etc.) can therefore serve as work site preparation tools, drivers of joint discussions between groups of stakeholders, risk analysis support and diagnosis tools, work site situation diagnosis and solution identification tools, etc. Furthermore, these means of visualising real-life work site situations may be used when training designers, in conjunction with situational simulations and experiences reported by experienced designers and prevention experts. Visualisation tools could be combined with storytelling techniques enabling designers to identify with workers so that they can project themselves into their "actual work".

---

[1] We changed the technical name for confidentiality purposes.

9.4.3 3D Models and Augmented Reality: Visualising for Action and Training Purposes

3D diagrams and 3D print models were identified as good practices which may be developed more systematically. The 3D model should not however be construed as covering all risks, as it must also be updated, and other observation scales may be required. For the RC46 work site for example, the modelling of the pillbox must feature cable runs. These temporary power cables may not have been taken into account during the environment "scanning" phase, even though they pose a potential threat.

As with many tools, the models proposed herewith should not be regarded by the organisation as a substitute for human labour or cooperation and consultation between stakeholders. On the contrary, simulation must enable the development of "discussion forums" on potentially hazardous situations and technical or organisational measures which can be put in place. For training purposes, these tools must be used to show the situation to the engineers assigned to design projects.

9.5 Discussion

Our data gives us a clear understanding of the visualisation issues involved in the organisational construction of occupational safety and security, in the specific case of very large-scale maintenance operations. These revamping sites mobilise national engineering teams, several production sites, numerous partner companies, project teams, etc. over many years. The central theme is to ensure the sustainable reliability of risk analyses, which must be transmitted between several groups of stakeholders forming separate "communities of practice" [4, 6, 20, 34]. One of the major difficulties is to raise the designers' awareness of the importance of taking worker safety into account from the project design phase, while they are still in the process of drawing functional and technical diagrams. At this stage in the project, details of concrete site working conditions are often unclear. Historically, designers tend to consider that the workers' safety will be managed "on the work site" and is therefore not their responsibility. This is a twofold challenge: raise their awareness of the importance of this issue and of their potential role in this respect, but also provide them with the tools they need to visualise the reality of workplace *situations*.

Care must therefore be taken not to emphasise the role of visuals, because the complexity of actual work may always greater than the image reflected by a model or video for example. Furthermore, the benefit of visualisation should not obscure the crucial and irreplaceable importance of human expertise. As mentioned by an interviewee, "someone who is not familiar with the equipment or risks may not take good photos of work sites, as they will be unaware of high-risk situations or hazardous materials". The simple choice of camera angle is significant: for example, a hazard may be linked to the cramped nature of a room, more so than the equipment in this room. In this example, a relevant photograph should seek a wide angle to show the work area rather than the equipment (pump, valve, motor).

Consequently, this is a mediation by the visualisation tool of an essentially human and organisational process, which begins with the recognition of the problem experienced by others, the awareness of its impact on the rest of the series, and continues with the desire to implement dialogue between stakeholders and related trade communities. The safety building process is above all human and organisational, or even "political" insofar as the groups of stakeholders involved can also have immediate positions and interests, i.e., directly compatible. Visualising work site situations helps make the risk tangible, concrete and directly assessable.

9.6 Conclusion

The study combines the characteristics and issues of high-risk industry, the challenges of occupational health and safety, the key role of design (prevention concepts derived from the BTP and *Prevention through Design* [1, 17] but also of cooperation and decision-making over the long term (management of projects and large-scale projects [2, 10, 14, 30, 32]) and the short term (notions of *sensemaking* [33], situation [7, 15, 29], etc.).

We attempted to combine this theoretical input with the empirical data collected to stress the importance of developing skills relating to "situated safety" for the company, i.e., the pragmatic understanding of occupational safety issues, generating interactions between legal constraints and processes on the one hand, and between the reality and specific characteristics of work sites on the other. These situated safety skills exist within the company and are very valuable. Thus, the resources in possession of these skills must be identified (often because of a dual work site/ safety culture) and the organisational and managerial conditions required for their enhancement must be created. This enhancement can only occur on a sufficiently local scale so that the transfer of knowledge is directly connected with action, work, the "investigation" and solving of actual problems, in conjunction and co-creation with designers.

These visualisation, three-dimensional modelling or virtual reality tools may therefore be used during "action learning" sessions intended for designers newly assigned to this position (more traditional training) as well as more experienced designers (sessions more oriented towards group work and the "co-design" of operational solutions, combining the expertise of prevention experts with the knowledge of designers). Finally, it should be noted that visualisation tools cannot be a substitute for the expertise of prevention experts and trainers familiar with the reality of work sites. Similarly, tools such as photographic databases, videos and 3D models, augmented reality and virtual reality, should not be regarded as self-sufficient, but rather as a means of grounding the activity and mediating regulation and cooperation among different yet complementary communities of practice.

9.7   Ethics Statement

Informed consent was obtained from all participants in this study, and all data has been anonymised. The research protocol was approved by a manager in the research division of the company.

References

1. D. Abowitz, M. Toole, Studying prevention through design: sociology meets civil engineering. Faculty Colloquia (2017)
2. A. Badri, A. Gbodossou, S. Nadeau, Occupational health and safety risks: towards the integration into project management. Saf. Sci. **50**, 190–198 (2012). https://doi.org/10.1016/j.ssci. 2011.08.008
3. T. Baxendale, O. Jones, Construction design and management safety regulations in practice—progress on implementation. Int. J. Project Manage. **18**, 33–40 (2000). https://doi.org/10.1016/S0263-7863(98)00066-0
4. B.A. Bechky, Sharing meaning across occupational communities: the transformation of understanding on a production floor. Organ. Sci. **14**, 312–330 (2003). https://doi.org/10.2307/413 5139
5. P. Béguin, Y. Clot, L'action située dans le développement de l'activité, Activites (2004)
6. J.S. Brown, P. Duguid, Organizational learning and communities-of-practice: toward a unified view of working, learning, and innovation. Organ. Sci. **2**, 40–57 (1991)
7. M. Charvolin, M. Duchet, Conception des lieux et des situations de travail. INRS Editions (2006)
8. F. Darses, Résolution collective des problèmes de conception. Le travail humain **72**, 43–59 (2009). https://doi.org/10.3917/th.721.0043
9. C. De La Garza, *Gestions individuelles et collectives du danger et du risque dans la maintenance d'infrastructures ferroviaires (PhD thesis in ergonomics)* (École pratique des hautes études, Paris, 1995)
10. M.K. Di Marco, P. Alin, J.E. Taylor, Exploring negotiation through boundary objects in global design project networks. Proj. Manage. J. **43**, 24–39 (2012). https://doi.org/10.1002/pmj.21273
11. E. Fadier, L'intégration des facteurs humains à la conception, travaux actuels et perspectives. Phoebus 59–66 (1998)
12. E. Fadier, C. De la Garza, Safety design: towards a new philosophy. Saf. Sci. Saf. Des. **44**, 55–73 (2006)
13. E. Fadier, C. De La Garza, A. Didelot, Safe design and human activity: construction of a theoretical framework from an analysis of a printing sector. Saf. Sci. **41**, 759–789 (2003). https://doi.org/10.1016/S0925-7535(02)00022-X
14. B. Flyvbjerg, M.S. Holm, S.L. Buhl, Underestimating costs in public works projects: error or lie? (2002)
15. J. Forrierre, F. Anceaux, J. Cegarra, F. Six, L'activité des conducteurs de travaux sur les chantiers de construction: ordonnancement et supervision d'une situation dynamique. Le travail humain **74**, 283 (2011). https://doi.org/10.3917/th.743.04
16. J. Gambatese, J. Hinze, C. Haas, Tool to design for construction worker safety. J. Archit. Eng. **3**, 32–41 (1997)
17. J.A. Gambatese, A.G. Gibb, B. Charlotte, T. Nicholas, Motivation for prevention through design: experiential perspectives and practice. Pract. Period. Struct. Des. Constr. **22**, 04017017 (2017)
18. C.D. la Garza, E. Fadier, Towards proactive safety in design: a comparison of safety integration approaches in two design processes. Cogn. Tech. Work **7**, 51–62 (2005). https://doi.org/10. 1007/s10111-005-0173-7

19. S. Gherardi, *Organizational Knowledge: The Texture of Workplace Learning* (Blackwell Pub., Malden, MA, 2005)
20. S. Gherardi, D. Nicolini, The organizational learning of safety in communities of practice. J. Manag. Inq. **9**, 7–18 (2000)
21. S. Gherardi, A. Strati, *Learning and Knowing in Practice-Based Studies* (Edward Elgar Publishing Ltd., Cheltenham, U.K, Northampton, Mass, 2012)
22. M. Grosjean, L'awareness à l'épreuve des activités dans les centres de coordination, Activites (2005)
23. C. Grusenmeyer, Interactions maintenance—exploitation en sécurité: étude exploratoire. Cahiers des notes documentaires. Hygiène et sécurité au travail **186**, 52–66 (2002)
24. N. Heddad, L'espace de l'activité, de l'analyse à la conception, PhD thesis in psychology, CNAM (2016)
25. N. Heddad, L'espace de l'activité: une construction conjointe de l'activité et de l'espace. Le travail humain **80**, 207–233 (2017)
26. INRS, Conception des lieux et des situations de travail. Dossier complet (No. ED 937). INRS, Paris (2005)
27. B. Journé, N. Raulet-Croset, La décision comme activité managériale située. Revue française de gestion N° **225**, 109–128 (2012)
28. B. Journé, N. Raulet-Croset, Le concept de situation : contribution à l'analyse de l'activité managériale en contextes d'ambiguïté et d'incertitude. Management **11**, 27–55 (2008)
29. F. Lamonde, J.-G. Richard, L. Langlois, J. Dallaire, A. Vinet, La prise en compte des situations de travail dans les projets de conception—La pratique des concepteurs et des opérations impliqués dans un projet conjoint entre un donneur d'ouvrage et une firme de génie conseil [WWW Document] (2010). https://www.irsst.qc.ca/-publication-irsst-la-prise-en-compte-des-situations-de-travail-dans-les-projets-de-conception-la-pratique-des-concepteurs-et-des-ope rations-impliques-dans-un-projet-r-636.html. Accessed 29 Aug 2012
30. C. Midler, "Projectification" of the firm: the Renault case. Scand. J. Manag. **11**, 363–375 (1995)
31. L. Rognin, I. Grimaud, E. Hoffman, K. Zeghal, Impact of delegation of spacing tasks on safety issues. Eurocontrol Exp. Centre, Bretigny, France (2002)
32. S. Spangenberg, Large construction projects and injury prevention, Doctoral dissertation (Dr.Techn). Denmark & University of Aalborg, Denmark (2010)
33. K.E. Weick, *Sensemaking in Organizations* (Sage Publications, T. Oaks, 1995)
34. E. Wenger, *Communities of Practice: Learning, Meaning, and Identity* (Cambridge University Press, Cambridge, UK, New York, NY, 1998)
35. M. Wolff, J.-M. Burkhardt, C. De la Garza, Analyse exploratoire de "points de vue": une contribution pour outiller les processus de conception. Le travail humain **68**, 253–286 (2005)

Open Access This chapter is licensed under the terms of the Creative Commons Attribution 4.0 International License (http://creativecommons.org/licenses/by/4.0/), which permits use, sharing, adaptation, distribution and reproduction in any medium or format, as long as you give appropriate credit to the original author(s) and the source, provide a link to the Creative Commons license and indicate if changes were made.

The images or other third party material in this chapter are included in the chapter's Creative Commons license, unless indicated otherwise in a credit line to the material. If material is not included in the chapter's Creative Commons license and your intended use is not permitted by statutory regulation or exceeds the permitted use, you will need to obtain permission directly from the copyright holder.

Chapter 10
Network Visualisation in Supply Chain Quality and Safety Assurance of a Nuclear Power Plant Construction Project

Kaupo Viitanen and Teemu Reiman

Abstract Supply chain quality and safety assurance aims to proactively create and maintain prerequisites for nuclear safety in the supply chain. An important task is being aware of the structure of the entire supply chain and how it affects safety. In this chapter, the authors describe how a network visualisation method was developed and used in supply chain quality and safety assurance of a nuclear power plant construction project.

Keywords Network visualisation · Nuclear power plant · Safety assurance · Supply chain

10.1 Introduction

Supply chain safety and quality assurance aims to proactively create and maintain prerequisites for nuclear safety in supply chains. An important task is being aware of the supply chain structure and how it affects nuclear safety. In megaprojects such as nuclear new builds, this is not trivial because the number of suppliers can be very high. For instance, approximately 2000 subcontractors were involved in the Finnish Olkiluoto 3 nuclear power plant construction project, reaching up to five tiers at the construction site [1].

In this chapter, the authors describe how a network visualisation method was developed and used to support supply chain quality and safety assurance of a nuclear power plant construction project. The benefits and limitations of applying this type of visualised representation of the supply chain for safety practice are discussed.

K. Viitanen (✉) · T. Reiman
Fennovoima Oy, VTT Technical Research Centre of Finland Ltd, Espoo, Finland
e-mail: Kaupo.Viitanen@vtt.fi

T. Reiman
Lilikoi, Helsinki, Finland

© The Author(s) 2023
J.-C. Le Coze and T. Reiman (eds.), *Visualising Safety, an Exploration*,
SpringerBriefs in Safety Management, https://doi.org/10.1007/978-3-031-33786-4_10

10.2   Visualising Safety and Network Visualisation

In safety science and practice, visualisations serve many purposes. At least four types of safety visualisations can be distinguished. Conceptual safety visualisations present some aspect of the concept of safety in a visual manner. They aim to answer the question "what is safety". Such visualisations are often educational and include a theory of how accidents occur, or what phenomena can influence safety. Examples of famous conceptual visualisations include Reason's Swiss cheese model [2] and Rasmussen's sociotechnical risk management and migration models [3].

Data visualisations convey safety-related data to end-users for easier recognition of patterns, for summarisation, or for economical communication of the data. They aim to answer the question "is it safe". Examples of safety-related data visualisations include conventional charts (e.g., bar, pie or line charts) of adverse outcomes, and their trends.

Visual tools help make sense of safety-related information. They involve user interaction, including inputting, organising and analysing data visually. Such tools aim to answer the questions "how does this relate to safety" or "is this safety". Examples of visual tools include the accident analysis methods ACCIMAP, FRAM and bowties.

Visualisations can also be used to communicate safety-related phenomenon in a dramatic and vivid manner, aiming to influence the viewers through creating affective responses. They answer the (possibly unasked) question of "what is safety" by explaining "this is safety" or "this is not safety" through means of narrative and dramaturgy. Examples of visual dramatisations include posters, movies, videos and websites with graphic content of accidents or their causation.

This chapter focuses on one type of visualisation process, the visualisation of networks. Network analysis examines the relationships between entities (e.g., friendship, communication or acquaintance networks). Although perhaps most commonly used for social networks, network analysis is not limited to social entities or phenomena, but it can be used with any relational data. In the context of supply chains, network analysis has been used for modelling contractual relationships, material flows, communication of instructions between the companies, performance incentives, etc.

Visualisation is an integral part of network analysis, because it facilitates, for instance, the detection of interactions and emergent patterns, and understanding the overall structure of the network. The most common visual representations of networks are node-link diagrams. Node-link diagrams consist of nodes, links that connect the nodes, and a layout (incl. node positioning and link routing). To improve the readability or emphasise some aspects of the diagram, various metrics are calculated based on network topology or the underlying data and are mapped to visual parameters of the diagram (e.g., node sizes and colours, and link colours and widths).

10.3   Supply Chain Network Visualisation Method

Fennovoima (the future operator of Hanhikivi-1 nuclear power plant) has granted an EPC (engineering, procurement, and construction) contract for a complete turnkey delivery of the power plant to the plant supplier, who in turn has made several contracts with vendors. According to Finnish legislation and regulatory requirements, the licensee is responsible for ensuring the safety of the nuclear power plant in all its life cycle phases [4]. In the context of supply chain quality and safety assurance, one of the implications of this requirement is that the licensee must have an overview of the status of the supply chain. The supply chain network visualisation method was developed as a partial solution to this issue. Its purpose was to help make sense of the project's contractual structure and support quality and safety assurance activities.

Before the network visualisation, information regarding the contractual relationships was in spreadsheets. The spreadsheets were rather complicated and hard to make sense of due to the sheer number of contracts and companies involved. Simplified visualisations that only described the most important top suppliers were also available, but they only contained a small fraction of the whole supply chain. An overall visualisation was not a high-priority task because each technical discipline was well aware of the companies that were directly connected to their job. However, for supply chain quality and safety assurance, a holistic perspective that takes the whole supply chain in account is necessary to understand how companies interact, where they are located in relation to other companies, and how they contribute to the overall safety of the construction project (and ultimately, the nuclear power plant). Network visualisation was chosen to provide this overview due to the following reasons:

- It provides an overview of contractual structure of the construction project.
- It provides a great deal of information at a glance without having to access the raw data (incl. safety classifications, contract grades and contract expiration dates).

During the years 2017–2019, the supply chain network visualisation has been produced six times (see Fig. 10.1 for example graph). In each update, new data was added and improvements were made to its visual design. It has evolved into a highly customised graph in response to practitioner needs. After two years of utilisation and development, the supply chain network visualisation method has established itself at Fennovoima. It has been incorporated into management system procedures as one of the methods periodically utilised for gaining an overview of the supply chain. The visualisation is still in continuous development.

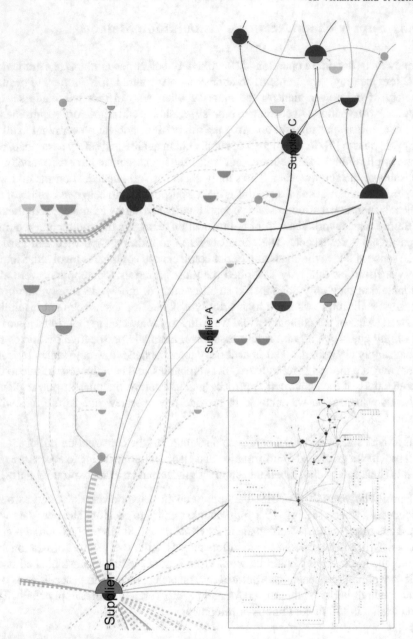

Fig. 10.1 Extract of the supply chain network visualisation (all company names have been removed and a few random modifications have been made to ensure confidentiality). Nodes indicate suppliers and links indicate contracts. Colours are mapped to contract safety classifications, e.g., red arcs refer to safety class 1 contracts (incl. reactor pressure vessel or primary circuit components). Node shapes indicate the structural positions of suppliers (left side indicates number of incoming contracts and right side outgoing contracts)

10.4   Application

The supply chain network visualisation has been applied as a decision support tool in defining the extent of supply chain quality and safety assurance activities, especially in the field of safety culture. As the future operator of the Hanhikivi-1 power plant, Fennovoima is responsible for assuring safety culture during construction. Fennovoima's safety culture assurance activities include monitoring activities such as audits, facilitation activities such as trainings and collaborative activities such as work groups with suppliers (for further details, see [5]).

In the nuclear industry, graded approach is applied to ensure that the application of requirements and the stringency of control measures is commensurate with nuclear safety significance. Fennovoima has developed a specific graded approach for supply chain safety culture assurance (SCA grade) to define safety culture assurance activities for a given supplier or sub-supplier on a four-level scale (A-D).

Fennovoima identified the need for safety culture specific grading when the supply chain grew in size and the supplier and the sub-suppliers signed multiple safety-classified contracts. Actors in the supply chain became more distant (due to increase in tier length) and their significance and interrelations became more difficult to comprehend. To maintain focus on significant suppliers, SCA grading process was developed as part of a wider development of supply chain safety culture assurance. SCA grades are reviewed and updated on an annual basis (for further details, see [5]). The initial determination of the SCA grade is based on the safety classification of the contract and the initial grade given by Fennovoima's supply chain management. Additional factors influencing the SCA grade include the type of contract and the position of the company in the project network. The supply chain network visualisation contributes to the grading process by identifying the position and role of each supplier in the overall project network and in relation to other suppliers. This approach borrows from social network analysis. The visualisation provides a way of positioning the supplier in the Hanhikivi-1 project and has already on a couple of occasions acted as evidence for raising the SCA grade of a particular supplier.

For example, the visualisation shows that Supplier A (Fig. 10.1) has multiple incoming safety-classified contracts (marked as red, orange and blue), but no outgoing ones. Hence, Supplier A acts as a key node in terms of delivering safety–critical services to multiple other companies. This position means that there are several companies already auditing and monitoring Supplier A, and providing information regarding its status. That is, a lot of information on this supplier is probably already available. This may include audit and inspection reports, observations, and other documented data. Consequently, Fennovoima needs fewer monitoring activities of its own. For example, Fennovoima does not necessarily need to do safety culture audit to Supplier A (or audit frequency can be decreased), if the customers of Supplier A have well-functioning auditing programmes. However, Fennovoima would first need to verify the capability of the customers of Supplier A to produce usable and reliable information, for instance by observing their audits, crosschecking their findings, or reviewing their assessment or inspection processes.

The position of Supplier A also suggests that there is a risk of common cause failure, because one supplier delivers to multiple customers. If this supplier fails to deliver an acceptable product, it can have widespread effects on the project. This has implications of quality and safety assurance of Supplier A: it might be necessary to investigate, what is the capability of Supplier A to manage multiple deliveries, with potentially differing, overlapping and contradicting requirements.

Supplier B (Fig. 10.1) is in a different position in the network. This supplier has a vast number of contracts with varying safety significance going out, and only one safety-classified contract coming in. It represents a key node in terms of oversight and contract management activities, including the distribution of requirements further down the supply chain and ensuring that sub-suppliers understand and apply them. Being in such a position suggests that Supplier B has an effect on many companies and needs to have a highly developed supply chain management practices of its own. However, many contracts of Supplier B are not categorised nuclear safety-significant (marked as green in Fig. 10.1).

The implications of the Supplier B position of the network include that there might not be as much documented information available from other companies in the network. However, due to the vast size of its supply chain, there exists a lot of knowledge in the sub-suppliers about Supplier B. This means that supply chain quality and safety assurance activities towards Supplier B (and other companies in similar positions in the network) are based on Fennovoima's own data collection and generally ensuring continuous and close collaboration practices with this supplier and its sub-suppliers.

Supplier C (Fig. 10.1) represents yet another type of position in the network. It has incoming as well as outgoing contracts. Supplier C sets requirements to others and delivers services based on requirements set by someone else. That is, its customers monitor it, but it also has the responsibility to establish and monitor its own supply chain. From Fennovoima's perspective, this company is not only responsible for quality products, but also for other suppliers. This calls for an assurance approach combining those described for Supplier A and Supplier B.

One of the main insights of the network visualisation has been in illustrating the networked nature of the project supply chain, and that suppliers in different positions require different supply chain quality and safety assurance approaches. In principle—and in hindsight—many of these observations could be deduced from the supply chain contract register spreadsheet without visual aids. However, the visualised representation of the supply chain proved helpful in orienting the supply chain quality and safety assurance activities to consider the positional relations between the suppliers, which is something that the tabular data was not able to do. Spreadsheet data is still needed when details or specific contracts need to be reviewed—as of now, the visualisation is too coarse a method for examining details. This suggests that it should be considered as a complementary tool among other tools in making sense of the supply chain.

10.5   Discussion

Experts from various disciplines at Fennovoima have communicated and presented the supply chain network visualisation in various events and meetings, ranging from top management meetings to nuclear safety committees and regulator inspections. The visualisation has been used to communicate the overall structure of Hanhikivi-1 project network, or to illustrate the reasoning behind supply chain assurance decisions. Overall, the reception and feedback has been very positive. The visualisation has been described as providing a good overview, or as offering a holistic picture of the supply chain in a simple way.

What does the supply chain network visualisation tell that non-visual data cannot? The application of the visualisation indicated that there are clear benefits to using visualised as opposed to tabular or textual descriptions of the supply chain.

First, it is an intuitive way of analysing complex data and phenomena. The visualisation helped identify patterns in the contractual data and connections between different suppliers, helping experts determine the suppliers' roles in the project network and consequently support designing quality and safety assurance strategies.

Second, it serves as an economical communicational aid for situations where time and efficiency is of the essence. Visual inspection of the diagram combined with a few examples of the different nodes gives the viewers an overview as they are able to see the entire network at a glance. Describing the basic functionality (e.g., colours, node sizes) of the graph, the viewers learn to read it quickly, which is not easily achieved with complex spreadsheets. This benefit was evident in many top-level meetings, where only a short time window was available for presenting. Using the visualisation, experts can present massive amounts of information.

Third, it is relatively easy to return to in later communications because the viewers are already familiar with the visualisation. Hence, it provides a memorable reference point for supply chain-related discussions or for sense making.

The visualisation might not be always useful for everyone. The end user's familiarity with the supply chain influences how they perceive the visualisation. For example, supply chain experts who know the underlying data intimately, and know how to read the supply chain contract register effectively, are likely to benefit less from the visualisation in terms of understanding how different companies relate to each other, or where they are visually located in the network because they already intuitively know this. This might not be the case for other experts or the management, who probably only know the suppliers most relevant to their tasks, but not the big picture. For them, the visualisation provides an easy-to-approach overview of the supply chain, something that tabular contract register is not able to provide.

While most viewers of the visualisation perceived it to be "interesting" and thought-provoking, its meaning or relevance—especially its relation to nuclear safety—was sometimes hard to grasp. In its current form, the only explicitly safety-related data included in the visualisation is the safety classification of contracts. Other safety-related information inferred from the visualisation relies on the experts' interpretation. For example, social network metaphor was applied to interpret the visualisation when assigning SCA grades, as described in the previous section. Adding more data points, especially ones that relate to safety (e.g., audit or inspection findings, etc.) may be a potential approach to make the visualisation more readily interpretable for end-users and to provide a more complete overview of the status of the supply chain. However, there are major drawbacks to adding more data. Increased visual clutter is the most critical one. Even in its current state the visualisation can be very hard to read in some areas, despite the efforts to make it clear. Trade-off between readability and amount of safety-related information included in the visualisation must be successfully managed. This means finding the answers to the following questions:

- What is the minimum amount of safety-related information that is required for the visualisation to be useful to safety practitioners?
- What is the minimum level of readability required that the visualisation would still make sense to the end-users?

Another solution to making the connection between the network visualisation and safety more evident is integrating a conceptualisation of safety into the diagram itself. That is, combining a data visualisation with a conceptual visualisation. To the authors' knowledge, there are no established (visual) safety models that explicate the connection between safety and supplier roles in contractual networks. For instance, Reason's Swiss cheese model is quite clearly focused on the operations within a single organisation and describing different types of barriers, and while Rasmussen's sociotechnical risk management model does, in principle, include external actors such as regulators or the government as part of the sociotechnical system, it still does not specifically address supplier organisations nor their roles. Neither model readily addresses how contractual structures (or other network phenomena) influence safety. This may suggest that there is a need for a completely new type of safety model, or a creative variation of an existing one.

A potential drawback of integrating a safety model with actual data is that the viewers may become anchored to this particular representation of data and perceive that the data and the safety model are inherently linked. The risk is that such a visualisation might become treated as an end-all solution to making sense of safety in the supply chain. In actuality, a visualisation (or in this case, a visual safety model) only projects the data in one way. Data in itself can have any number of projections. Similar process occurs implicitly when the viewers of the current version of the network visualisation attempt to relate the visualisation to safety by applying their mental models of safety. Therefore, integrating conceptual models of safety in the visualisation requires care from its developers, and informed practitioners who know the assumptions underlying the model and are able to avoid the safety model having too much (or unwanted) influence on their thinking or decision-making.

The supply chain network visualisation was sometimes observed to induce affective responses on experts or decision-makers: the colourful visualisation was perceived as attractive, or its complexity was perceived as shocking. Integrating (or at least considering) an explicit safety model may help better manage such effects. It is important that visual dramaturgy directs attention to the right things, and to the ones most relevant to safety. Ideally, an engaging visualisation makes the various stakeholders more aware of the importance of managing quality and safety in the supply chain and the challenges it involves, and be more motivated and committed towards solving them. One of the risks is that the visualisation conveys the supply chain as too complex in a too simple way. That is, the viewer might only remember the visual complexity or its (potentially attractive) visual appearance (i.e., the things that caused the affective response), and not the safety-related insights embedded in the visualisation (*cf.* picture superiority effect). Therefore, knowledge in designing visual narratives or generally visual storytelling is important to understand what kind of responses the visualisation creates in viewers to manage its dramatic impact.

10.6   Conclusions

In this chapter, the authors described a supply chain network visualisation method, its background and an example of its application in a nuclear power plant construction project. The method was developed as a solution to better make sense of how a project network creates preconditions for safety. Experiences showed that the visual representation of the supply chain helped uncover such insights of the supply chain that probably would have remained hidden if relying on tabular data only. These insights were applied in designing supply chain quality and safety assurance activities. Further development needs were also identified, especially the development of safety models that explicitly address the safety significance of various interaction phenomena in safety–critical networks.

References

1. P. Oedewald, N. Gotcheva, Safety culture and subcontractor network governance in a complex safety critical project. Reliab. Eng. Syst. Saf. **141**, 106–114 (2015). https://doi.org/10.1016/j. ress.2015.03.016
2. J. Reason, *Managing the Risks of Organizational Accidents*, 1st edn. (Ashgate, Aldershot, Hants, England; Brookfield, Vt., USA, 1997)
3. J. Rasmussen, Risk management in a dynamic society: a modelling problem. Saf. Sci. **27**(2–3), 183–213 (1997). https://doi.org/10.1016/S0925-7535(97)00052-0
4. STUK, *Radiation and Nuclear Safety Authority Regulation on the Safety of a Nuclear Power Plant.* (Säteilyturvakeskus, Helsinki, Finland, 2018)
5. T. Reiman, K. Viitanen, Safety culture assurance in the supply chain of a NPP construction project. Presented at the probabilistic safety assessment and management conference, Los Angeles, USA (2018)

Open Access This chapter is licensed under the terms of the Creative Commons Attribution 4.0 International License (http://creativecommons.org/licenses/by/4.0/), which permits use, sharing, adaptation, distribution and reproduction in any medium or format, as long as you give appropriate credit to the original author(s) and the source, provide a link to the Creative Commons license and indicate if changes were made.

The images or other third party material in this chapter are included in the chapter's Creative Commons license, unless indicated otherwise in a credit line to the material. If material is not included in the chapter's Creative Commons license and your intended use is not permitted by statutory regulation or exceeds the permitted use, you will need to obtain permission directly from the copyright holder.

Chapter 11
Visualization for the Safe Occupation of Workspaces

Elsa Gisquet and Gwenaële Rot

Abstract Through the example of the underground construction site for the extension of the Parisian subway, it will be analysed how, in this fluctuating, dark environment, visualization help to maintain safety requirements. How does visualization help with work? Based on the observation that pictorial representation can be used to drive and organize activities, this chapter will highlight the ways in which these visual artefacts advance safety in three ways, by helping participants inhabit, discuss, and synchronize their workspaces.

Keywords Safety · Occupational hazards · Risk management · Human safety environment

11.1 Introduction

An important literature on system safety theory focuses on technical and organizational phenomena found in high-reliability organizations (HROs). The following observations of industrial settings typical of HRO conceptual profiles and system properties were based on intensive case studies of large-scale organizations with very stable infrastructure. The empirical industrial field is not identical across nuclear power plant operations, businesses, and manufacturing situations, and so far this approach has not been replicated in other types of settings where fluctuating infrastructure gives rise to changing workspaces.

Underground infrastructures are not specifically dedicated and adapted to production: they are at once workplaces and work materials (extraction). Space below ground is defined by techniques, practices, and values, both an "arena"—that is, a physical space that constrains actions—and a "setting" in the sense that it can be

E. Gisquet (✉)
Institut de Radioprotection et de Sûreté Nucléaire, Cherbourg-en-Cotentin, France
e-mail: elsa.gisquet@irsn.fr

G. Rot
Centre de Sociologie des Organisations, Sciences Po, Paris, France

© The Author(s) 2023
J.-C. Le Coze and T. Reiman (eds.), *Visualising Safety, an Exploration*,
SpringerBriefs in Safety Management, https://doi.org/10.1007/978-3-031-33786-4_11

rearranged materially or at least mentally by the individuals working in it, depending on their needs [13].

In this fluctuating, dark environment, to what extent does visualization help to maintain safety requirements? How does visualization help with work?

This case study of the extension of a Paris underground metro line shows that pictorial representations play a daily role as a mobile and moral authority in the infinite reconfiguration of space. Based on the observation that pictorial representation can be used to drive and organize activities, this chapter will highlight the ways in which these visual artefacts advance safety in three ways, by helping participants inhabit, discuss, and synchronize their workspaces.

11.2   Inhabiting the Space

The case of the metro line construction site combines activities taking place within a restricted perimeter, both above and below ground. As underground boundaries shift, those on the surface adapt. Ongoing territorial conflicts require that the borders of the workspace be constantly re-specified.

Updating representations through various maps and diagrams is therefore a constant necessity to help workers to identify potentially risky areas of coactivity, including the arrival of new hazards. This work of updating representations also requires organizational labour, an ongoing process that comes together as the work is being carried out. This necessitates a constant exchange of information, between graphics designed to represent a given phase of the work and the configuration of the work in situ as it is observed over the course of different worksite visits.

Maps, including the coordination maps among different stakeholders required by regulation, appear as "boundary objects" [16] in the management of coactivity.

These boundaries may become the subject of clashes over definition among different work teams, as a means of appropriating spaces in order better to adapt them to the demands of their work. In these cases, the maps bear the markings of the work teams' adjustments to and clashes over these definitions (Fig. 11.1).

At stake is more than workers' ability to locate themselves spatially: they must also appropriate and adopt this space. Workers must inhabit the space both in their bodies and in their perceptions [5]. This inhabiting spans from individual experience to collective (family, group) management dynamics [4]. In this case, perception of the work space is not limited to the relation of the body to the machine [7, 18], but more broadly of the body to its environment. Pictorial representations help to make workspaces into familiar environments by integrating their codes and prohibitions.

Visual artefacts offered by management may aid in prevention actions, some of which may be seen as governing the conduct of operators (Fig. 11.2). This term refers to the capacity of visual representations to shape, guide or influence people's behaviour. Here, the meaning of the word "conduct" goes beyond the idea of imposed direction, referring also to the way in which an individual behaves when guided by a sense of self-regulation [8].

Fig. 11.1 Site plans annotated to identify borders and hazards in work space

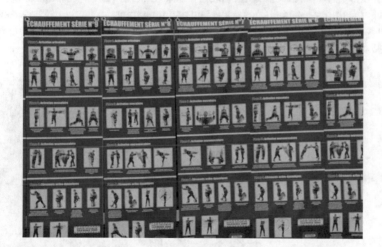

Fig. 11.2 Visual aid for management recommendations

Beyond that, visual artefacts can also help workers to appropriate a space in order to improve their own attention to safety (Fig. 11.3). In addition to rules and formal descriptions, little notes written by workers make it possible to alert colleagues that certain points require special attention or to specify particular ways of doing things [3, 14].

This appropriation of space by workers [10] is not only professional. Inscriptions, graffiti, etc. help to make workspaces into familiar spaces, they contribute to collective effort while helping to minimize stresses that may discourage new workers or even drive them away [15].

Fig. 11.3 Little notes help
appropriate workspaces in
the service of safety

   Visual artefacts are also vectors of identity and pride in workspaces, helping to
create a sense of permanence or territory (Fig. 11.4). It is not uncommon for workers
to take pictures of their work, while others show pictures of previous work sites they
are proud to have participated in. Interest and pride in the profession is a central
element of stability—sometimes conveyed and maintained by a family environment
[1]. Mutual acquaintance and attention paid to others is another contributing factor
to safety [1, 9, 15].

Fig. 11.4 Pride in the work
done

The narrow white margin in this schedule shows which activities will immediately delay the rest of the schedule (the pink margin) if they run late; i.e., which delays would cause the opening of the metro line to be delayed.

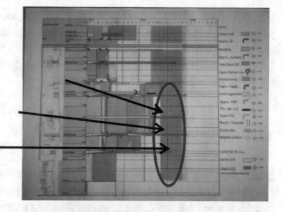

Fig. 11.5 Strategic spaces identified by a visual representation of the schedule

11.3 Discussing the Space

11.3.1 Locating Strategic Space

The work of updating spaces is also a work of organizing activities as they are being carried out. It implies ongoing forward and backward movement between maps and graphics and the configuration of activity on the ground.

For example, a theoretical work schedule for a construction site developed and proposed during a competitive bidding process does not necessarily correspond to the actual work schedule once the building is actually underway [17]. Schedules do not account for the unexpected, the implication being that unexpected occurrences are linked to failure or diversions and therefore necessarily generated by others [6]. Local adjustments must take place after the space is discussed and debated to better organize it. Visual artefacts help to identify which deadlines must be met for the metro to be operational on time (Fig. 11.5).

11.3.2 Local Regulation

Companies must succeed in inhabiting spaces not only in fact but in speech, given the performative of the latter. For this reason, coordination meetings are a platform for debating the space–time of work, using visual artefacts as justification or support. These artefacts serve as a traceability instrument in support of speeches and other arguments. For example, if a company claims to have cleared a workspace before leaving, coordinators can use visual evidence to contest that claim.

Sketches and other diagrams, whether drawn up on a white board or scribbled on the back of an envelope, make it possible to consider as a collective the overall dynamic process of operations and the flow of activities. They make it possible to link techniques to organizations, which improves safety [12].

Visual artefacts help to build a narrative of one's workspace, and at the same time, bolster the reliability of operations by improving their ability to adapt and adjust in the field, especially in high-tech contexts.

11.4 Synchronizing the Space

Situations such as the one studied here are characterized by a continuous flow of changes; operators inhabiting these territories transform them into spaces whose boundaries are not fixed, but always subject to transformations and re-actualizations [6]. These transformations do not operate according to the same logic. While the overall focus at the site is on the metro line under construction, this focus is supported by micro-movements and adjustments that occur over different types of time (Fig. 11.6): the rational chronological time expressed in the planning phase, unexpected time, political time (the metro must be operational by a certain date for the sake of the public), local residents' time (they don't want to be disturbed over the weekend while at the same time they want to be able to use the metro as soon as possible), the time of companies that want to meet market demands to ensure profitability.

These temporalities and temporal perceptions [11] are different for different groups of people, and these dyscronies [2] make it complicated to organize the flow of activities in workspaces, creating regulatory deficits and making it difficult to contain a task and its hazards in a single, controllable space–time.

Fig. 11.6 Different spatio-temporal frames on the same site

This being the case, visual artefacts facilitate temporal synchronization, allowing for better fluidity of activities and the coordination of work spaces. Drawings, diagrams and sketches help to build bridges among different temporalities, creating alignment between human and non-human elements, between field activities and Office requirements.

Coordination meetings are practical situations that make it possible to identify, propose, record, and formalize any deviation from the planned-for model as the construction work progresses, in particular situations of coactivity. In this context, visual artefacts function as traceability instruments, as a "memory" of the construction site, understood as a series of photographs taken as the site's conditions and stakes evolve.

At a minimum, visual artefacts are instruments that make it possible to produce both a temporality and a memory for a given site. Visual artefacts are part of local site regulation [14].

11.5   Conclusion

Maintaining safety in spaces that are constantly reconfiguring requires constant updating and reorganization, which is aided by visual artefacts.

There is a vast and varied literature on human reliability, ranging from human factors engineering to socio-technical systems. This literature has already highlighted that high-reliability organizations must constantly achieve extraordinary levels of operational reliability, while working constantly to improve it.

In this context, visual representations of workspaces offer an opportunity to analyse and organize workspaces, even as they succeed each other and change over time.

For the researcher, at least, these visual artefacts shed light on the practice of these territories: their movements, direction, and temporality; how operators inhabit these territories and transform them into spaces whose boundaries are not fixed, but always subject to transformation and re-actualization [6].

References

1. I. Aldeghi, V. Cohen-Scali, Orientation et professionnalisation des jeunes dans le secteur du bâtiment. Cahier de Recherche (219)
2. N. Alter, Mouvement et dyschronies dans les organisations. L'Année sociologique 53(2), 489–514 (2003)
3. A. Borzeix, B. Fraenkel, Langage et travail: communication, cognition, action, CNRS éd. (CNRS communication, Paris, 2001), 369p.
4. P. Bourdieu, Le sens pratique. Actes de la recherche en sciences sociales 2(1), 43–86 (1976)
5. M. Breviglieri, Habiter l'espace de travail. Perspectives sur la routine. Histoire et société. Revue Européenne d'Histoire Sociale (9), 18–29 (2003)

6. M. De Certeau, *L'invention du quotidien, 1. Arts de faire* (Éditions Gallimard, Paris, 1990)
7. N. Dodier, Les hommes et les machines: la conscience collective dans les sociétés technicisées, Editions Métailié (1995)
8. M. Foucault, A.I. Davidson et al., *The birth of biopolitics: lectures at the Collège de France, 1978–1979* (Springer, 2008)
9. C. Gaudart, C. Delgoulet, et al., La fidélisation de nouveaux dans une entreprise du BTP. Approche ergonomique des enjeux et des déterminants. Activités **5**(5–2) (2008)
10. E. Goffman, *Relations in Public* (Transaction Publishers, 2009)
11. J. Hassard, Images of time in work and organization, in *Studying Organization: Theory and Method*, pp. 327–344 (1999)
12. T.R. LaPorte, P.M. Consolini, Working in practice but not in theory: theoretical challenges of "high-reliability organizations". J. Public Adm. Res. Theory: J-PART, pp. 19–48 (1991)
13. J. Lave, *Cognition in Practice: Mind, Mathematics and Culture in Everyday Life* (Cambridge University Press, 1988)
14. G. Rot, A. Borzeix et al., Introduction. Ce que les écrits font au travail. Sociologie du Travail **56**(1), 4–15 (2014)
15. M. Santos, M. Lacomblez, Que fait la peur d'apprendre dans la zone prochaine de développement? Activités **4**(4–2) (2007)
16. S.L. Star, J.R. Griesemer, Institutional ecology, translations and boundary objects: amateurs and professionals in Berkeley's Museum of Vertebrate Zoology, 1907–39. Soc. Stud. Sci. **19**(3), 387–420 (1989)
17. S. Tillement, J. Hayes, Maintenance schedules as boundary objects for improved organizational reliability. Cogn. Technol. Work, pp 1–19 (2019)
18. P. Ughetto, Les nouvelles sociologies du travail: introduction à la sociologie de l'activité, De Boeck Superieur (2018)

Open Access This chapter is licensed under the terms of the Creative Commons Attribution 4.0 International License (http://creativecommons.org/licenses/by/4.0/), which permits use, sharing, adaptation, distribution and reproduction in any medium or format, as long as you give appropriate credit to the original author(s) and the source, provide a link to the Creative Commons license and indicate if changes were made.

The images or other third party material in this chapter are included in the chapter's Creative Commons license, unless indicated otherwise in a credit line to the material. If material is not included in the chapter's Creative Commons license and your intended use is not permitted by statutory regulation or exceeds the permitted use, you will need to obtain permission directly from the copyright holder.

Chapter 12
Screening Workplace Disaster: The Case of *Only the Brave* (2017)

Shane M. Dixon and Tim Gawley

Abstract Media influence how we define and engage with our world, shaping our interpretations, attitudes, behaviours. Feature films in which work-related injuries, deaths, and disasters are the storylines can convey occupational safety messages to large, diverse audiences. Films can entertain, act as "powerful" and "poignant" memorials to workers, heighten peoples' awareness of events, and even deepen their understanding of the causes of workplace disasters. However, it is unclear how films actually represent the complexities of workplace injury and industrial disaster. We examined the film *Only the Brave* (di Bonaventura, Luckinbill (Producers), Kosinski (Director) in Only the Brave [Motion Picture] (Columbia Pictures, United States, 2017)), which recounts the story of the deaths of 19 wildland firefighters in America. In particular, we examine how the film portrays workplace disaster and the factors which led up to the event. We discuss some strengths and limitations of feature films as a form of visualizing workplace disaster.

Keywords Workplace injury · Disaster · Risk · Film · Media representation

12.1 Introduction

The Yarnell Hill Fire in Arizona, United States of America (USA), began with a lightning strike on 28 June 2013 [3]. The fire consumed approximately 8300 acres of land, destroyed 114 structures, and forced the evacuation of thousands of people, before it was contained on July 10th [1]. Among the assets mobilized to contain the fire were the "Granite Mountain Hotshots," a highly trained Interagency Hotshot Crew who are tasked with the most challenging, critical assignments involved in fighting wildland fires in the USA [17]. On 30 June 2013, the Yarnell Hill wildfire

S. M. Dixon (✉)
Leadership Program, Wilfrid Laurier University, Waterloo, ON, Canada
e-mail: sdixon@wlu.ca

T. Gawley
Department of Health Studies, Wilfrid Laurier University, Waterloo, ON, Canada

© The Author(s) 2023
J.-C. Le Coze and T. Reiman (eds.), *Visualising Safety, an Exploration,*
SpringerBriefs in Safety Management, https://doi.org/10.1007/978-3-031-33786-4_12

killed 19 of 20 firefighters from this crew: it is the highest death toll of US firefighters in a single event since 11 September 2001.

The feature film, *Only the Brave* (2017), follows the story of the Granite Mountain Hotshots over the course of several years as they fight wildfires in America's southwestern states and ultimately to the fire that entrapped and killed them. The focus on workplace disaster is rare among major motion pictures but similar to recent movies such as *Deepwater Horizon* [4]. *Only the Brave* provides an opportunity to examine a re-telling of a story about a multi-casualty disaster in a high-risk context.

Representations of workplace disaster, which can be shared in many different forms—such as news articles, documentaries, feature films, and government reports—can teach us about how events occur and how to prevent them [4, 5, 7]. Film *can* visually communicate the story of workplace disaster, message how it should be seen, and provide insights into its causes. Importantly, when we look at any visualization of health and safety we also need to bear in mind that while these are ways of seeing "their downside is that they are ways of not seeing" [11], p. 80. Story compositions, and the "narrative choices" that go into them, influence what audience members will see and how they will understand the causes of workplace disaster [13]. Films are organized to form a particular narrative that includes (and excludes) information and emphasizes some factors while deprioritizing others. As such, part of examining film as visualizations of safety is to identify *what is* and *what is not* present on the screen.

12.2   Communicating Safety in Feature Film

There is a well-developed literature on the role that films play in shaping culture and understandings of the world. It points to film's power to change views, reinforce perspectives, educate, and/or to be used as a consciousness-raising mechanism [16]. Feature films have distinct characteristics such as large budgets and casts of celebrity actors. They appeal to emotions and are visually exciting. Hollywood films are intended to earn a large return on investment and therefore are designed to entertain the largest audiences possible. They often do this by creating dramatic content and using well-worn conventions [10, 16]. A cinematic story with selfless heroes performing extraordinary acts, overmatched by forces seemingly beyond control, and against a backdrop of spectacular special effects, is a successful and perennially used template. Some of these same characteristics privilege seeing workplace disasters in one way rather than another such that more micro-level factors are emphasized.

Only the Brave displays a degree of what we refer to as *perspectival alignment* between (a) individualistic safety perspectives (e.g., human error, "person approach" [14]) that focus on the safe or unsafe actions (or inactions) of individuals in a workplace that create the conditions that cause injury or disaster [8], and (b) the content and form of Hollywood feature film (e.g., well-worn tropes, closed story arcs, 90–120 min length). Renditions of disaster stories that can be presented in an exciting, non-complex way (i.e., limited number of actors, short span of history, clear heroes and villains), with clear cause-and-effect relationships and in approximately two hours, fit extremely well with film industry needs. American film tends to focus on individuals (or small groups) and individualistic attempts to prevent or mitigate a disaster [4]. These films concentrate on the immediate causes of disaster such as ignored safety warnings, failure to follow safety protocols and the agency of the heroic characters, while deemphasizing or excluding the larger context and its influence. In doing so, they exclude questions such as "why were safety protocols not followed?" and "what conditions preceded violating safety protocols?" To be sure, immediate causes are integral parts of the story. However, left out, or at least minimized, are the structural- and organizational-level factors that are critical explanatory factors for the disaster (e.g., [8, 9]) whose inclusion would give viewers a more fulsome visualization of the causes of workplace disaster.

12.3   Wildland Firefighting Crews

Interagency Hotshot Crews (IHCs) or "hotshots" are groups consisting of 20 persons who are tasked with directly attacking a wildland fire. They receive a great deal of training; often hike long distances to reach fires (or are occasionally helicoptered to the fire), and fight them by building firelines around a fire that slow or stop its advance by starving it of fuel. For hotshots, building fireline involves physically removing fuel (e.g., trees, shrubs, grass) down to the mineral soil from the fire's path in a 1–1.5 m line around the fire. To remove combustible materials, hand tools such as Pulaskis, shovels, and chain saws are used. Hotshots also use "back burning," the practice of depriving a wildfire of fuel by burning sections of forest between a fireline and the advancing fire. In a combined effort to stop a fire, hotshots often work with other firefighters such as engine crews (fire trucks) and aerial suppression crews (e.g., planes and helicopters). All of these techniques and tools are used in the film.

12.4   Analysis

12.4.1   A High-Competence Crew

Only the Brave makes a very persuasive case that the Granite Mountain Hot Shots were a highly competent crew in a high-risk context (see [2] regarding wildland fire-fighting competence). We identify four distinct, but non-exclusive, competencies— *cognitive, leadership, technological,* and *physical*—that contribute to protecting the crew's health and safety which are portrayed in the film. The presence of each of these competencies support the narrative that the crew had the knowledge, skills, and attitudes essential to effectively and safely fight fire.

Cognitive competency refers to the formal and informal knowledge that firefighters acquire and practice about fire behaviour such as how it is affected by fuel types, weather and terrain, and how to suppress it. This enables crews to assess fire dangers, develop attack strategies, coordinate resources, and devise escape plans. The Granite Mountain Hotshots are not incautious or reckless. Multiple examples are shown in the film in which the crew receives a fire call and they quickly, but systematically, assess the size and behaviour of a fire, strategize an attack, and plan an escape route. They assess the conditions by communicating with incident command elements, other crews, reading topographical maps, closely monitoring the weather, and "reading the fire", looking at fuel types, prevailing winds, and how the fire is acting. When their plan is in place, the firefighters deploy to their various positions to construct fire line and protect themselves.

The crew's adeptness at "reading the fire", is on display in multiple scenes. Early in the film, the crew offers advice to another group of hotshots suggesting they control a fire in a particular way. This suggestion is rejected by the hotshot crew. In a scene that highlights the superintendent's knowledge of fire behaviour, we see the neighbourhood that the crew was trying to save engulfed in flames and the Captain points out, "They should've listened [to us], supe [superintendent Marsh]. Could've saved a lot of people's homes." This is on display again when the crew unconventionally but successfully contains part of an intensifying fire in steep terrain and once more in dramatic fashion when their well-constructed fireline and backburn halts the advance of a raging fire.

Leadership competency refers to the decision-making, problem-solving, and directional style of individuals in a group. Superintendent Marsh is portrayed as a firm but caring leader: he is willing to discipline the crew if they are not meeting the high standards he has set but nurturing as he provides lessons about fire behaviour and technical skills. In demanding, stressful situations he is decisive and the crew are respectful of him. While he is the clear leader, he asks for feedback from his crew.

Visuals of *technological competency* permeate the film. This competency refers to the use of physical technologies such as the measurement instruments and fire suppression tools. Throughout the film, hotshots skillfully use the tools and technologies including the use of hand tools such as axes and shovels, power tools such as chainsaws, and drip torches. We also see examples of firefighters being taught how to properly use the tools of their trade. For instance, the superintendent coaches a rookie firefighter, in the proper use of a drip torch as they light a "backburn" to eliminate vegetation, depriving the fire of fuel.

The strength and endurance that are essential for wildland firefighting, the *physical competency* of hotshots, is portrayed throughout *Only the Brave*. The crew intensely trains multiple times in the film, running for several kilometres and practicing digging line. The crew's physical abilities are also depicted throughout the film in scenes where they are hiking long distances with their equipment through steep, difficult-to-navigate terrain, digging line in hot, smoky conditions.

12.4.2 Hazard Identification

Hazards abound for wildland firefighters and include falling trees, uneven and steep terrain, the tools and equipment of fire control and suppression, and of course, fire itself. Wildfire is the obvious antagonist and the most threatening hazard of the film and is presented in visually stunning ways. When the fire is moving slowly, we see the crew work amidst the flames and smoke trying to control the fire's advance. In numerous breathtaking scenes we see towering walls of flame approach the crew that works unperturbed by its advance, toiling in smoke, embers, and firebrands. Enhancing the differences in power between the fire and crew, are wide angle camera shots that are used to place the firefighters in panoramic scenes working ahead of an encroaching blaze. These shots accentuate the smallness of the firefighters in a vast terrain: a small number of individuals armed only with rudimentary hand tools grossly outmatched against the huge wildfire.

Importantly, while there is omnipresent danger from fire throughout the film, it is simultaneously recognized as a force that can be safely controlled through the crew's expert capabilities and the proper prevention practices for safely fighting wildland fire.

12.4.3 Hazard Prevention

We see the firefighters relying primarily on two protections in the film. First, we see them using personal protective equipment (PPE) such as goggles, helmets, clothing, and footwear and we see them practicing the deployment of vital fire shelters, shielding that will protect from radiant heat but not direct flame should they be overrun by fire. The other type of protection that predominates in the film are administrative controls: rules, guidelines, and training about how to effectively and safely fight fire. In several scenes, we see the firefighters rely on their training and rules to decrease their exposure to hazards. For example, we see the crew enact proper procedure as they post lookouts, check weather readings, and watch for spot fires that have crossed their fireline. Among the procedures, explicitly referred to are the "10 Standard Firefighting Orders and 18 Situations that Shout Watch Out." These are the guidelines issued by the US Forestry Service and have been relied on for 40 years to protect firefighters (see [2]). The import of these is clearly demonstrated in a tension-filled scene during a training exercise when a firefighter forgets one of the "10 and 18" and the superintendent—inches from the rookie firefighter's face—growls, "firefighters died for us so we could learn all these Watch Outs"! The superintendent then forces the whole crew to do push ups as collective punishment for not remembering the safety rules.

12.4.4 Communication

The nature of wildland firefighting requires well-coordinated movement and communication between and within groups involved in fire suppression activities. *Only the Brave* displays how communication facilitates safe and effective fire management. The film depicts coordinated fire suppression activities between ground and aerial fire crews. Amidst plumes of dark, heavy smoke, and the rhythmic thumping of helicopter blades, the film opens with a scene that portrays strong coordination between air and ground units as a helicopter flies over the crew who directs the pilot precisely where to drop water on a fire. In several scenes, we see senior incident command elements, poring over topographical maps, planning fire suppression activities and passing direction on to subordinates. On the ground, communication among the Hotshots, whether it be through shouts, hand gestures, over radios, is illustrated as concise, practiced, and unerring even against the crackling of the fire, screaming of chainsaws, and the thudding and scratching of digging tools impacting earth.

As the film progresses, there are depictions of errors and the limits of technology that disrupt communication exposing crew members to hazards. While coordination among crew members remains strong, there are breakdowns in information exchange between the ground and air attack crews. During the "Dragon Fire" a large, four-engine air tanker drops its multi-tonne load of water on an incorrect target and a massive wall of water narrowly misses the crew. During the Yarnell Hill Fire scenes, a tanker mistakenly drops its water on the Hotshot's backburn, extinguishing it. Hampered by both smoke and terrain, the ground and air crews are unable to communicate with the Granite Mountain Hotshots before they are overtaken by the wind-driven, fast-moving wildfire [6].

12.4.5 Entrapment

Prior to their entrapment the Granite Mountain Hotshots were in a safe position—an area already burnt by the fire—had assessed the situation, considered it safe to move, and were moving in an organized way to a pre-planned safety zone. We cannot be "in the heads" of the crew so their true motivation for moving is unclear. Their route took them from atop a ridge where they could watch the fire into a box canyon where a ridge obstructed their view of the fire. At the same time, as the fire was approaching his position, the crew's look out relocated and he was unable to see and report on the fire. Both of these events restricted the crew's ability to adapt to the fire's behaviour. As the crew hiked to their new location, the fire, swept along by wind gusts and tinder-dry fuel intercepted them. The weather conditions were dynamic, the winds shifted direction and increased and fire intensified. When it became apparent their route was cut off by fire, the crew, following their training, cleared a deployment site and readied their fire shelters for the burnover. Simultaneously we see the communication breakdowns between ground and aerial crews, and desperate, unsuccessful attempts to reach them. Confined in the canyon, unable to be helped by ground or aerial firefighting units, the crew's position was overtaken by fire, killing the 19 hotshots.

12.5   Discussion

Only the Brave chronicles the work of a hotshot crew in a high-risk context. It presents the hazards they faced, the controls they used to prevent harm, and the crew's tragic deaths. Films such as *Only the Brave* can provide entertainment and increase people's awareness of disasters and their causes. In so doing, they render workplace disasters visible. Completely dismissing these films as only having entertainment value is not a good direction if we are interested in raising awareness of, and enhancing knowledge about, workplace disasters. However, *Only the Brave* tells a particular story. It focuses on individuals' behaviours and largely follows typical templates of Hollywood feature film. As such, audiences are presented with a rendering of the story that focuses heavily on the crew's actions and minimizes, and in some cases excludes, the role of structural and organizational factors.

Only the Brave uses a standard cinematic story template that focuses on a single group of characters at the micro-level of workplace disaster. We are shown the crew's competence, cohesiveness, expertise in employing firefighting practices, and their adherence to the Standard Firefighting Orders. The film tightly "zooms in" to present the immediate causes of the disaster: the crew's decision-making, the changing weather, in particular the wind, and the fire. This parsimonious set of factors provides a close look at the circumstances leading up to the entrapment but in doing so leaves out the structural, cultural, and organizational factors. These are critical factors to explore in any disaster story (e.g., [9]). There is much we do not see antecedent to entrapment, the canyon, and the Yarnell Hill Fire itself.

"Zooming out" and exploring distal causes or in the words of Reason [14], "latent conditions," is more challenging as they are not considered in the film. The exclusion of the distal causes is, in Rae's [13] words a "narrative choice", which is partly driven by the film industry's need for easy-to-understand, exciting, and closed-ended cinematic stories. As Quinlan [12] states, "while unsafe practices in workplaces cannot be ignored, focusing on them in isolation from the social structures and interests that encourage them is misleading and ineffectual" (p. 169). Unfortunately, the film does not present viewers with the latent conditions that influenced the disaster such as organizational factors and the role of these are left unexplored. Several critical factors contributed to the disaster [18] but are omitted from the film. Questions about whether the crew should have been fighting the fire or whether incident command should have deployed them where they were in light of the extreme fire conditions—high winds, drought conditions, and an abundance of fuel—are not problematized. Indeed, the area that the hotshots were fighting fire was in drought conditions and had not burned in 47 years, creating conditions for extreme fire intensity known to be difficult, if not impossible to control. A larger, though related question concerns the efficacy of trying to protect structures in the urban-wildland interface when these are deemed "indefensible" [18]. Any discussion about whether "complexity of the operation exceeded the organizational capacity of the firefighting system mobilized to respond to the fire", including whether there were sufficient firefighting resources and inclusion of all relevant stakeholders in decision-making ([6], pp. 43–44) and the part these played in the disaster is also left out. The influence of the US Forestry Service's practice of "aggressively fighting all wildfires" and relative risks to firefighters' safety is also not represented [15]. Dixon and Gawley [4] found a similar lack of attention to the broader context in the film *Deepwater Horizon*.

These portrayals have implications for the use of films for understanding workplace disaster. *Only the Brave* can reproduce the notion that despite precautions, safety systems will fail, and there is nothing that can be done by workers or managers. Such a notion has the potential to forestall consideration of the influence of latent conditions not presented in the film (e.g., managerial decision-making). Another notion that is reproduced in *Only the Brave* is the "person approach" [14]: the cause of the disaster was the workers' unsafe decision-making. Looking only at the film one could be forgiven for thinking the hotshots, in haste or hubris wrongly and unsafely decided to reposition. Settling on this explanation places sole responsibility on workers, negating the macro factors and latent conditions that influence individuals in the local context. Without considering the influence of cultural and organizational factors we are left, unfortunately, with blaming the "incompetent dead" [2].

12.6 Conclusion

The spectacular nature of films, such as *Only the Brave*, draws in viewers and highlights the hazards to which workers are exposed, raising awareness and enhancing understanding about workplace disasters. Importantly, while stories about disaster are screened, through filmmakers' narrative choices particular parts of the story are screened out, ultimately hindering our understanding of the disaster. The film industry's need for spectacular stories influences how disasters are covered and how organizational factors, for example, contribute to the disaster are minimized or excluded. This can influence peoples' understandings of workplace injury and disaster. *Only the Brave's* focus on the firefighters' abilities and actions in the lead up to the entrapment, without attention to factors such as firefighting doctrine and managerial decisions and how these intersect with natural environmental conditions, serve to reproduce a person approach narrative without raising questions about the role of socio-political and organizational factors. The educational value of these films can increase with greater integration of individual *and* structural elements, providing a more comprehensive picture of workplace disaster.

References

1. R. Anglen, D. Wagner, Y. Wingett Sanchez, New account of hotshot deaths in Yarnell fire. *USA Today* (2015, April 10). Retrieved from https://www.usatoday.com
2. M. Desmond, *On the Fireline. Living and Dying with Wildland Firefighters* (University of Chicago Press, Chicago, 2007)
3. L. di Bonaventura, T. Luckinbill, T. Luckinbill, (Producers), J. Kosinski (Director), *Only the Brave* [Motion Picture] (Columbia Pictures, United States, 2017)
4. S.M. Dixon, T. Gawley, Crude exploration: portraying industrial disaster in *Deepwater Horizon*, a film directed by Peter Berg, 2016. New Solut. **27**, 264–272 (2017)
5. T. Gawley, S.M. Dixon, "Covered in Coal": examining workplace injury, illness and fatality in *Blood on the Mountain*. New Solut. (2020). https://doi.org/10.1177/1048291120907306
6. K. Hardy, L.K. Comfort, Dynamic decision processes in complex, high-risk operations: the Yarnell Hill Fire, June 30, 2013. Saf. Sci. **71**, 39–47 (2015)
7. J. Hayes, S. Maslen, Knowing stories that matter: learning for effective safety decision-making. J. Risk Res. **18**, 714–726 (2015)
8. A. Hopkins, What are we to make of safe behaviour programs? Saf. Sci. **44**(2006), 583–597 (2006)
9. A. Hopkins, *Disastrous Decisions: The Human and Organisational Causes of the Gulf of Mexico Blowout* (Lightning Source Inc., 2012)
10. A. Kersten, M. Verboord, Dimensions of conventionality and innovation in film: the cultural classification of blockbusters, award winners, and critics' favourites. Cult. Sociol. **8**, 3–24 (2014)
11. J.-C. Le Coze, Safety, Model, Culture: The Visual Side of Safety, in *Safety Cultures, Safety Models: Taking Stock and Moving Forward*. ed. by C. Gilbert, B. Journé, H. Laroche, C. Bieder (Springer, Cham, 2018), pp.81–92
12. M. Quinlan, *Ten Pathways to Death and Disaster: Learning from Fatal Incidents in Mines and Other High-Hazard Workplaces* (The Federation Press, Leichhardt, 2014)
13. A. Rae, Tales of disaster: the role of accident storytelling in safety teaching. Cogn. Technol. Work **18**, 1–10 (2016)
14. J. Reason, Human error: models and management. Br. Med. J. **20**(7237), 768–770 (2000)
15. M.P. Thompson, D.G. MacGregor, C.J. Dunn, D.E. Calkin, J. Phipps, Rethinking the wildland fire management system. J. Forest. **116**(4), 382–390 (2018)
16. G. Turner, *Film as Social Practice*, 4th edn. (Routledge, New York, 2006)
17. United States Forest Service, *Hotshots* (2020). Retrieved from https://www.fs.fed.us/science-technology/fire/people/hotshots
18. Wildland Fire Associates, *Granite Mountain IHC Entrapment and Burnover Investigation*. Arizona Division of Occupational Safety and Health (2013). http://www.npstc.org/download.jsp?tableId=37&column=217&id=2944

Open Access This chapter is licensed under the terms of the Creative Commons Attribution 4.0 International License (http://creativecommons.org/licenses/by/4.0/), which permits use, sharing, adaptation, distribution and reproduction in any medium or format, as long as you give appropriate credit to the original author(s) and the source, provide a link to the Creative Commons license and indicate if changes were made.

The images or other third party material in this chapter are included in the chapter's Creative Commons license, unless indicated otherwise in a credit line to the material. If material is not included in the chapter's Creative Commons license and your intended use is not permitted by statutory regulation or exceeds the permitted use, you will need to obtain permission directly from the copyright holder.

Chapter 13
Post-script: Visualising Safety

Teemu Reiman and Jean-Christophe Le Coze

Abstract This concluding chapter on visualisation for, and of, safety, weaves together ideas put forward by the volume's contributors. It analyses how visualisations and their role have changed over time, their co-evolution with key concepts in safety science and impact on cognitive representations built by practitioners, whether success should be assessed by level of adoption or by impacts on safety outcomes. A number of open questions for future research are outlined.

Keywords Visualisation · Usability · Evaluation · Safety science · Communication

13.1 Introduction

The workshop was a success in terms of the quality of the original contributions and the debates ranging from the usability of different types of visualisation to the status of safety science in general. An interrogation on visualisation can trigger fundamental questions about our relationship with the world. It is far from trivial, and investigating visualisation has multiple implications.

In this respect, we included a wide variety of potential questions and topics to explore into the original call for papers reproduced in the introduction of this book. This was done on purpose. The workshop was to be an exploration of a previously under-researched topic. We did not expect an answer to all, or even most, of the questions. We did expect vivid discussions about the topic of visualising, together with some tentative answers to our questions combined with ideas and directions for future research. In this we succeeded. Next, we provide a brief summary of answers received to the questions we posed in the Call.

T. Reiman
Lilikoi, Lohja, Finland
e-mail: teemu.reiman@lilikoi.fi

J.-C. Le Coze (✉)
Ineris, Verneuil-en-Halatte, France
e-mail: jean-christophe.lecoze@ineris.fr

© The Author(s) 2023
J.-C. Le Coze and T. Reiman (eds.), *Visualising Safety, an Exploration*,
SpringerBriefs in Safety Management, https://doi.org/10.1007/978-3-031-33786-4_13

13.2 Visualisations of the Past, Evolution and Successful Examples

First questions we asked concerned issues such as examples of successful visualisations in safety, the evolution of visualisations over time and ways of classifying the diversity of existing visual artefacts.

Many contributions touched this topic. It became clear that whether a visualisation can be judged to be successful depends on how success is defined. If "successful" refers to effectiveness in terms of contributing to safety, there is little empirical research to make any conclusions although presentation of empirical case studies of a design project by Stoessel and Racuna (Chap. 9) or of daily operations in the construction phase of an underground by Gisquet and Rot (Chap. 11) make it clear how visuals matter in practice (one could also refer to Flach's contribution on ecological design, Chap. 8).

Many contributions showed how visualisations of concepts such as Swiss cheese and the Sharp-End Blunt-End (see especially Chaps. 4 and 5) have been influential among researchers and practitioners. This is the second challenge in answering the question: Some visualisations are accepted and shared among the safety community much more than others, making these "successful" in terms of their spread and utilisation. If we measure success by prevalence of use, posters in general can be considered among the very successful safety visualisations. However, modern safety posters differ from the safety posters of the past. As Swuste et al. (Chap. 2) illustrated in their contribution, the underlying ideas presented in posters have changed during the years. Yet there is little empirical research on whether effectiveness of posters in safety prevention has improved.

The same challenge of defining a successful visualisation exists when we look at some of the widely used models or methods. It has been noted that the usability of the method is more important than scientific concerns ([6], see also Chap. 4) in facilitating its acceptance among the practitioners. Again, how useful the widely accepted methods and models are in accident prevention, when compared to other less widely used methods and models, remains an unanswered question.

Visualisations, especially those that become widely accepted, also change through time. The evolution of visualisations question was addressed by Swuste et al. (Chap. 2), Waterson (Chap. 4) and Haavik (Chap. 5), especially in relation to posters, metaphors and concepts. However, further characterisation of the evolution of visualisation in safety is a potential future research topic. This would include the question of how the underlying representations of safety and accidents contribute to different types of visualisations and how visualisations in turn can have an effect on these representations (a co-evolution of representations).

13.3 Participants' Own Experiences in Visualisation

The second set of questions in the Call concerned practical experience of using visualisations in research and practice. We asked the participants to either reflect on their own research and the role of visualisation in it, or to take a look at how practitioners and/or researchers produce, use and disseminate diverse visualisations in their daily activity. These topics are clear future research areas as none of the contributions directly addressed them. However, there were many indirect implications and ideas in the many of the contributions. The topic was also discussed at length during the workshop, as all participants had practical experience in using visualisations in either research or practice of safety.

Even if personal experience of using visualisations was somewhat lacking in the contributions, some examples based on past studies were presented. For example, some contributions of the workshop showed how researchers have produced and reproduced visualisations, and how this process gradually changed both the visuals and the underlying representations (see also Waterson's work on the development of Accimaps e.g., [7]).

There are many interesting research avenues to pursue in relation to this topic. For example, what can the use of visualisations by practitioners reveal about safety as a practice? Can we learn something about safety practice by studying visualisations that are used by practitioners, and how practitioners use and re-use safety visualisations, whether of their own making or borrowed from safety literature?

13.4 Visualisations and the Science of Safety

The third set of questions in the Call focused on the role of visualising in conceptualising safety as a scientific concept. The questions concerned the contribution of visualisations to the framing of safety as a scientific object and how visualising a concept can change the concept.

These questions were again highlighted in many of the contributions and further discussed during the workshop. When discussing the evolution of visualisations over time (see above), it was noted that the underlying representation may also change. This co-evolution of the concept itself and the visualisation of the concept is an interesting future research topic.

The role of visualisations in the creative process of safety science is interesting. Looking at history (see especially Chaps. 2, 4 and 5), visualisations seem to play a major part in theory development in safety science, and maybe in science in general. However, some interesting questions warrant further attention: How much do visualisations merely communicate what the researcher is trying to convey, and how much do they also help the researcher to conceptualise his/her ideas? Are visualisations used only after the theory or model has been conceptualised, or in parallel? And how about the influence that the visualisation has on further development of the underlying representation?

James Reason's model of organisational accidents (the so-called Swiss cheese model) is an interesting example in illustrating the role of visualisation in model development, since Reason did not himself use the metaphor of a Swiss cheese. He developed a model together with John Wreathall of defence in depth applied to organisations and accidents, and represented this with slices with holes in his 1990 book Human Error. It is only later that an acquaintance, Rob Lee, came up with the cheese metaphor [4]. This invention by a colleague clearly influenced Reason, since his later visualisations resembled Swiss cheese much more than the earlier ones. In fact, to go one step back in history, Reason's original model was based on the metaphor of a human body and its resident pathogens combining with external factors to bring about disease. Reason was inspired by the human body analogy while using another analogy from the nuclear industry brought by John Wreathall. It is interesting to consider how much the underlying model was changed when the metaphor of Swiss cheese was invented, or when the metaphor was communicated to researchers and practitioners who all adapted the metaphor to their particular context of use. Certainly, the model can be considered among (the most) successful visualisations in safety science, and the strong debates following the validity of the model demonstrate its influence on science and practice of safety. Whether the influence has been bigger than warranted by the merits of the underlying model itself brings us to the final topic of possibilities and limits of visualising.

13.5   Possibilities and Limits of Visualising

The fourth set of questions in the Call asked about the possibilities and the limits of visualisation. The specific questions concerned issues such as the dangers of visualising complex phenomenon such as safety, what kind of opportunities new technology offers, the role of big data, and how videos or movies portray safety. We were also interested in the possible biases created by visualisations as well as how visualisations guide the attention of public and experts.

The contribution by Dixon and Gawley (Chap. 12) provided insight into the question concerning movies. What storylines emphasise and what they leave in the background characterise the situated perspective of the movie director. The pros and cons of a narrative approach to visualisation was also discussed at length during the workshop. This is connected to the issue of "selling" the visualisation to its intended audience. This selling can be done by different means, one of which is dramatisation. Another selling tactic can be simplification of complex phenomena. Further tactics could be humour, or otherwise clear and distinguishing visualisation. One could argue that Swiss cheese was successful also in this regard. We return to this topic in the concluding section of this chapter.

The possibilities and limits of visualising were discussed at length during the entire workshop, with the insight from John Flach's extensive experience with the design of ecological interfaces which guide their users in operational contexts [1, 2]. It was noted that visualising is always situated in a historical context. Visualisations have been created for a certain purpose in a certain context. Sometimes the purpose has not been clear even for the visualiser, and sometimes the purpose cannot be met (sometimes there are multiple, partly conflicting purposes). All visualisations, no matter when (or how, or why) they have been created, are always interpreted in the context of their current use. Visualisations are thus always (more or less) fit for purpose, and knowing what this purpose was is important for subsequent use of the visualisation.

The visualisation will emphasise some aspects of the phenomenon of interest, while obfuscating other aspects. However, it never dictates how the user eventually perceives the representation underlying the visualisation. Thus, instead of asking how a certain visualisation contributes to safety, we can ask what lines of reasoning different visualisations support. Then our focus is on the possibilities and constraints that a certain visualisation imposes on the user, including what aspects of the visualised phenomenon are emphasised and what aspects are de-emphasised.

The choice of what to leave out in visualisation is as crucial as is the choice of what to include. This can be studied from the perspective of the one doing the visualisation: how visualisations are created, and how conscious is the process of selecting the issues to include or emphasise and issues to leave out or de-emphasise? Are there some typical issues that safety visualisations under-represent?

13.6   The Way Forward and New Questions

Another future research topic could be the dark side of visualisation: misrepresentation, misuse and dramatisation in visualisation. Visualisation offers a way of highlighting issues of interest in a way that captures the attention of the perceiver. This added freedom of imaging also creates possibilities to misrepresent issues differently from text, for example. Misrepresentation can be accidental (e.g., due to lack of safety knowledge) or intentional (e.g., part of an organisational attitude change campaign or propaganda by interest groups). However, one should not underestimate the perceiver and his or her ability to see behind the surface of the visualisation: what have been the motives of the designer of the visualisation. A bad visualisation can in fact reinforce the opposing message.

The above issues bring forth an interesting question: Who is making the visualisations and what do they know (or should know) about safety (managers, designers, human factors experts, communications specialists, movie directors)? Future research could aim at clarifying the limits and possibilities of co-creation between safety experts (whether scientists or practitioners) and other interested parties such as designers, safety managers or communications specialists.

Another potential future research area is the role of trade-offs in visualising. As highlighted many times during the workshop, visualisations always emphasise some aspects to the detriment of other aspects. In addition to the question of what to visualise, the designer of the visualisation needs to balance between various other tensions; how much to simplify a complex phenomenon without oversimplifying it, how much to highlight (or dramatise) some aspects without distracting other important aspects, how much to explain (e.g., by text) and how much to leave for the perceiver to make sense.

One interesting research topic is the role of text in safety visualisations. During the workshop, it was noted that many of the posters presented by Swuste et al. (Chap. 2) were mostly composed of stylised text, whereas Castan's posters, as shown by Portelli et al. (Chap. 3), had pictures with little or no text. Humans process text differently from images, and this may affect how visualisations with or without text are understood. In many visualisations of safety models, text is typically used to explain signals or signs that can have multiple meanings, e.g., arrows. However, more "simple" or "information poor" visualisations, such as posters, often lack this explanatory imagery, making them more subject to multiple interpretations.

Finally, another topic is the relationship between visualisation and art. Have the heuristic, cognitive and enduring influences of some visualisations discussed in the workshop anything to do with their aesthetic dimensions? Posters, movies but also drawings clearly exhibit artistic features. Drawing is about selecting, ordering and combining shapes, lines, colours and sometimes texts to follow one's imagination when trying to make sense of something.

Painting is the same, whereas directing a movie is about composing with images, movement, lights, landscapes, sound and characters into stories something which is obviously deeply artistic. This connection is therefore another research area. Finally, art is deeply connected with imagination, as is visualising. What is the role of creativity and imagination in safety visualisation?

Many of the issues raised in the chapter deal with the wider issues of the place and use of safety visualisations in the context of development of safety science and practices. We hope this brief exploration of the topic of visualisation stimulates further studies, and also further visualisations for, and of, safety.

References

1. K.B. Bennett, J.M. Flach, *Display and Interface Design* (CRC Press, 2011)
2. J.M. Flach, The ecology of human-machine systems I: introduction. Ecol. Psychol. **2**, 191–205 (1990)
3. T. Haavik, Remoteness and sensework in harsh environments. Saf. Sci. **95**, 150–158 (2017)
4. J. Larouzée, J.-C. Le Coze, Good and bad reasons: the Swiss cheese model and its critics. Saf. Sci. **126**, 104660 (2020)
5. J. Reason, *A Life in Error: From Little Slips to Big Disasters* (Routledge, 2013)
6. P. Underwood, P.E. Waterson, Systemic accident analysis: examining the gap between research and practice. Accid. Anal. Prev. **55**, 154–164 (2013)
7. P.E. Waterson, D.P. Jenkins, P.M. Salmon, P. Underwood, 'Remixing Rasmussen': the evolution of Accimaps within systemic accident analysis. Appl. Ergon. **59**, Part B, 483–503 (2017)

Open Access This chapter is licensed under the terms of the Creative Commons Attribution 4.0 International License (http://creativecommons.org/licenses/by/4.0/), which permits use, sharing, adaptation, distribution and reproduction in any medium or format, as long as you give appropriate credit to the original author(s) and the source, provide a link to the Creative Commons license and indicate if changes were made.

The images or other third party material in this chapter are included in the chapter's Creative Commons license, unless indicated otherwise in a credit line to the material. If material is not included in the chapter's Creative Commons license and your intended use is not permitted by statutory regulation or exceeds the permitted use, you will need to obtain permission directly from the copyright holder.

Printed in the United States
by Baker & Taylor Publisher Services